dot COMPLICATED

Untangling Our Wired Lives

RANDI ZUCKERBERG

HarperOne
An Imprint of HarperCollins*Publishers*

HarperOne

DOT COMPLICATED: *Untangling Our Wired Lives.* Copyright © 2013 by Randi Zuckerberg. All rights reserved. Printed in the United States of America. No part of this book may be used or reproduced in any manner whatsoever without written permission except in the case of brief quotations embodied in critical articles and reviews. For information address Harper-Collins Publishers, 10 East 53rd Street, New York, NY 10022.

HarperCollins books may be purchased for educational, business, or sales promotional use. For information please e-mail the Special Markets Department at SPsales@harpercollins.com.

HarperCollins website: http://www.harpercollins.com

HarperCollins®, ≝®, and HarperOne™ are trademarks of HarperCollins Publishers.

FIRST EDITION

Designed by Level C

Library of Congress Cataloging-in-Publication Data
Zuckerberg, Randi.
Dot complicated : untangling our wired lives / Randi Zuckerberg.
p. cm.
ISBN 978–0–06–228514–0
ISBN 978–0–06–232710–9 (Intl)
1. Internet—Social aspects. 2. Information technology—Social aspects. I. Title.
HQ1178.Z83 2013
004—dc23   2013021760

13 14 15 16 17 RRD(H) 10 9 8 7 6 5 4 3 2 1

To my incredible husband, Brent Tworetzky,
for always being Dot Calm and Collected, and never
Dot Complicated. Thank you for being such a wonderful
father, a loving husband, and the world's best teammate.

CONTENTS

INTRODUCTION

You only live once. So make sure you spend 15 hours on the internet every day, desperately seeking the validation of strangers.

—@ChrisRock0z

For the past eight years I have had a front-row seat to how technology, mobile devices, and social media have changed, enhanced, and complicated almost every facet of our lives—from how we interact with our friends to how we elect presidents, from how we manage our careers to how we support the causes we're passionate about, from how we find love to how we raise our children.

I've seen the world radically shift from a place where connecting online with others was so new, so novel, so special, and where being reachable on a mobile device felt almost magical . . . to a world where we're now so connected online, so reachable, and so comfortable interacting with others from behind a screen, we often forget to look up and enjoy the world around us.

In the following pages, I take you through some of my own experiences, as I witnessed this remarkable shift firsthand, and share some of my personal observations on how social media has transformed the world we live in.

We have such powerful technology at our fingertips. But we need to make sure our attachment to being online doesn't get in the way of our lives and relationships offline. We need to find balance between being connected to millions of people around the world and being present with the people we love, standing right next to us.

It's complicated.

Over the past few years I've spent a lot of time thinking about the concept of "tech–life balance." What does that even mean? How do we find it? In an era where we are constantly connected, always on call, and reachable around the clock . . . is this kind of balance even possible?

But before I dive further into my thoughts on finding tech–life balance within our friendships, our families, our relationships, our careers, and our communities, I want to take you back a few years, to a time in my life that was anything but balanced. To a time where I worked tirelessly to help build one of the very tools that today presents us with so much joy and so much anxiety.

Maybe you're picking up this book because you and I are connected through social media. Maybe you've been a loyal Facebook user from day one. Maybe you don't have a clue who I am, but you think my brother is cool. Or maybe you have no idea why you're even reading this right now, but here you are. However these pages landed in your hands, thanks for joining me on this adventure. I am thrilled to have your attention, and I hope this book—part personal story, part thoughts for the future, and part guide for finding the right balance of tech in your life—can spark a dialogue around using technology mindfully, thoughtfully, and meaningfully to truly enhance our lives.

FIRST AND LAST STEPS

There are moments in life when everything changes.

Sometimes these moments come out of nowhere, ambushing you. Sometimes they approach from a distance and arrive so slowly and expectedly that change is nothing to be surprised about.

And then, sometimes, the moment comes when you open your mouth and blurt your heart out at the most random and surprising opportunity. That's how my life changed on April 20, 2011.

Sitting at my desk at work that morning, I had no idea this day would become one of the most important of my life and come to define my dreams, my career, and my views on technology and society in so many ways. But I already knew that it was going to be a very special day. Or at least a very strange one.

I had stopped by my desk for a quick breather. I don't normally start the day tired. Most of the time, I'm a morning person—one of those terribly perky people who are always ready to roll, even before their morning coffee. But I was thirty-five weeks pregnant at this stage, and the baby felt like it weighed fifty pounds. At a short five feet two inches, I was officially all stomach and none too sprightly on my feet.

I'd also been at work for about eighty hours—and things were only just getting started. My desk was a mess of call sheets, floor plans, and the wreckage of discarded takeout meals. I slumped in my chair and took a moment to collect myself.

Suddenly, the mess began to buzz. I fumbled wildly beneath the papers for my phone. I tapped the keypad and pressed it against my ear.

"*Randi!!!*" boomed an enthusiastic voice. "It's Ron!"

It was Ron Conway, the legendary Silicon Valley venture capitalist. He's a good friend, and I have all the time in the world for Ron on any occasion. Well, *almost* any occasion. This time I wilted slightly in my chair. It was too early, and I was too tired for such a rousing greeting.

"Hello, Ron," I said, forcing as much energy into my voice as I could, hoping I sounded less leaden than I felt. "What can I do for you?"

Ron paused for a moment. Then he spoke, as focused and earnest as he always is. "Listen, Randi. I need you to get M. C. Hammer in to see the president."

For a moment, my brain struggled to process the sheer unlikelihood and absurdity of what I'd just heard. And then I began to smile. Suddenly I felt a lot less tired.

The president's motorcade was on its way to Facebook. This was the day I'd been waiting for, the crescendo of my entire career to date.

The call had come in from the White House exactly two weeks earlier.

At Facebook, I was often contacted by people asking us to do events with them, sending movie scripts for our consideration, and pitching opportunities for our executives. Most of the time, these invitations weren't exactly right for us, and I was used to politely declining several dozen a day.

Then, out of the blue, the White House communications office phoned. They had seen some of my "Facebook Live" broadcasts and were wondering, would Facebook be interested in hosting President Obama for a town hall event in two weeks?

It's not every day that you get a call from the White House. So, even though I knew we would have to move mountains to pull this one off, even though all we had was an empty warehouse and a couple of cameras, I did the only sensible thing. I agreed on the spot.

Not only did the president want to come to Facebook to talk to our employees, he also wanted to answer questions submitted by people on the website as part of a "Facebook Live" event. The town hall would be streamed live, and people could tune in on Facebook to watch and ask questions. This was no idle PR stunt. The president was coming as part of his nationwide tour to make his case for a new economic policy—a strategy for cutting the deficit while maintaining investment in growth. For the president, it was a tough moment politically, and it wasn't clear whether he would prevail over congressional Republicans.

For Facebook, this was a defining moment. The president had every distribution channel available to him to communicate to the country. But out of every website, every TV channel, every radio station at his disposal, he had chosen Facebook as the best way to speak directly to the nation.

Power, drama, technology—all the makings of an epic marketing moment for Facebook. Even as I digested the basic details of the event, my brain was already racing through the possibilities of what we could do—and what we needed to do—to host the president at Facebook in only two weeks.

We worked nonstop for thirteen days. It was a crazy, exhilarating, and terrifying whirlwind in which each day blended into the next, and my team and I lived from conference call to conference call, meeting to meeting, espresso shot to Red Bull can (or for my pregnant self, decaf coffee to herbal tea). We had to sort out logistics and security with the White House—not something to mess around with. We had to work out how to market the event so that people on Facebook would know when and how to tune in. We had to decide who would get to attend, how we could collect questions, and whether we should try to focus the conversation at all.

We had a moderator—my brother, Mark Zuckerberg, Facebook's founder and CEO. But that's all we had when I hung up the phone on day one. We didn't even have a venue. There was only one space that we knew would fit the town hall. But it was nothing more than the big empty warehouse within our office complex. It had none of the furnishings or technical capacity to be a place where you'd host a presidential town hall.

We had to turn that warehouse into a fully functional studio and auditorium. We needed to find a crew to run the cameras. In fact, we needed cameras, lights, and audio. And we needed an Internet connection fast and stable enough to broadcast the president's message live to the world.

By the fourteenth day, I'd been working almost nonstop for eighty hours, with only the occasional break for a quick nap or bite to eat. The chairs had been set up, the security team had done their final sweep, and the cameras had been tested, and then tested again. I finally went home on the morning of the event to get presentable for the president's arrival. But this was no fancy-suit-and-slacks combo. This was Silicon Valley, and I was rocking my "on-camera Facebook signature look" of jeans with a custom-BeDazzled T-shirt, which displayed the Facebook logo in rhinestones.

And then it was straight back to work. My colleague Andrew Noyes, who had been instrumental in getting us to this day, met me at my house, just a few blocks from the Facebook office, to go over the plan for the day a final time and walk over to the event. I was glad for Andrew's company. Even though I hadn't slept in days, I was alert and energized, knowing how much was at stake.

Walking to the office, we saw multiple satellite trucks with antennas pointed to the sky, ready to send their signals far and wide. On the rooftops, the dark outlines of sniper rifles were barely visible. Everywhere there were security barricades, patrol cars, and lots of men clearly auditioning for parts in a Ray-Ban commercial. A police helicopter circled overhead, rotors thumping loudly.

I had stopped at my desk, simultaneously exhausted and energized, answering a few last-minute questions and staring out the window at the scene of managed chaos, when Ron called me.

"Listen, Randi. I need you to get M. C. Hammer in to see the president."

I smiled at the request. "Hold on a moment, Ron."

Pulling out the walkie-talkie attached at my waist, I radioed my colleagues at the ticket and credential area. "Malorie? Maureen? Come in. Can we arrange a seat for M. C. Hammer?"

A brief pause and then, "Roger that. We have secured a seat for M. C. Hammer."

I turned back to my phone. "Sure thing, Ron! Hammer's all set."

And then I was off to manage a thousand other last-minute crises. It was Hammertime.

. . .

A few hours later, I sat in my seat in the audience. The audience was hushed, expectant. We were waiting for the president to enter. Out of the corner of my eye, I spotted M. C. Hammer snapping photos with his cell phone.

I was desperately nervous. This was the moment we had poured so many hours of work into. Expectations were unnervingly high. If it went well, the victory would be sweet. If it went badly—well, I didn't want to think about that. There was too much on the line.

"So, when are you expecting?" whispered a voice.

Behind me sat a beaming Nancy Pelosi.

"Er, next month."

Nancy launched into an enthusiastic story about her grandkids. I like Nancy and had interviewed her earlier for the live stream as a warm-up to the president. But I felt distracted and left the talking to her. I wanted the event to start. The waiting was killing me.

And then the president arrived. People stood and applauded. The novelty of a sitting U.S. president coming to address a crowd of Silicon Valley's technorati was not lost on anyone.

Mark greeted the president, and they sat on two high stools, facing each other. The room grew quiet.

The president started. "Well, thank you so much, Facebook, for hosting this, first of all," he said. "My name is Barack Obama, and I'm the guy who got Mark Zuckerberg to wear a jacket and tie."

The room erupted. But I was doing everything I could just to stop myself from imploding. Against the odds, we'd done it, and it was going to be great. It had been an amazing journey to reach this moment.

Facebook was born in a dorm room. It grew up quickly, in unlikely circumstances, run first by a visionary team of students, and then with a growing crew of seasoned professionals attracted by a dream and an idea: that, by connecting people, we could give a voice to millions, transform the relationship between individuals and institutions, and help everyone get closer to the people who matter most to them. Every morning we would wake up and try to make the world more social, and we lived and breathed that mission every single day.

And now, incredibly, the president of the United States had come to Facebook to make the case for his agenda. For me, this was a pivotal career moment and one that was intensely personal. But there was also a much larger transformation taking place.

Mark was cool and collected. The president was forceful, charismatic, and energetic. There had been some doubt earlier in the day about whether he'd win the room. This was the low point of the president's first term, and a lot of his natural supporters were feeling the blues. At the event, those fears quickly vanished.

But he did more than just win the room. He had a bigger goal and a larger audience. The president had come to speak to *everyone on Facebook*. Even as he joked with Mark and focused intently on the different questioners, his eyes kept returning to the cameras dotted around the room. That was where the debate over America's economic strategy would be decided. In living rooms, offices, dorms, and coffee shops all across the country, people were tuning in to watch that town hall online. And he pulled no punches as he spoke to them.

"I know that some of you who might have been involved in the campaign or been energized back in 2008, you're frustrated. . . . Just remember that we've been through tougher times before. We've always come out ascendant. We've always come out on top. If we come together, we can solve all these problems. But I can't do it by myself."

People cheered. So did I.

But I wasn't just cheering for the president, as much as I agreed with him. I was cheering for those cameras—*my* cameras—that had just delivered the moment to the people outside this room. I was cheering for all my amazing colleagues who had worked with me to turn a dusty warehouse into a town hall fit for a president. I was cheering for the sheer audacity of what we had done: accepted an invite to run an event before we were even vaguely capable, and then made it possible through hard work and speedy improvisation.

Later, after the president and the crowds had long gone, I found out that it had been Facebook's biggest ever live-streaming event. The numbers were off the charts.

The tiredness I had felt that morning was gone entirely. I felt invigorated, flushed with success. All of us at Facebook had done something great with technology. Of course, I always thought that the engineers and product designers at Facebook were transforming the world. But this was tangible, immediate. We had used Facebook's enormous reach to enable a democratic experience for Americans and facilitate a political discussion for millions of people simultaneously.

That day the conversations at Facebook weren't about the insular struggles of Silicon Valley, the latest in-jokes among the engineering team, or even the next set of product milestones, as important as they were. People were talking about big issues. Inside and outside Facebook, we had created a moment that touched people.

I had now been working for nearly ninety hours straight. The set of the town hall was already being dismantled, and my team had taken off for the day. It was time to go home.

I walked slowly through the streets of Palo Alto. I didn't have my car. Driving was out of the question, because someone had closed all the streets.

The realization came slowly. *Oh. Wait. That was me.*

I was distracted, and as I walked, I felt incredibly restless. Today had been such a rush, such an amazing payoff after two weeks of nonstop work. But in that eruption of happiness, I felt a challenge mixed in with the reward. The gears were spinning furiously inside my head as I neared home. I walked up the path and climbed the steps to my front door. "Brent?" I called to my husband as I pushed open the door.

There was a rustling in the kitchen. A familiar furry face appeared from around the corner.

"Beast!" I cried. It was Mark's dog.

Another familiar face appeared. It was Mark. The tie and jacket from earlier, so unnatural on my brother, had vanished. The trademark hoodie and jeans were back.

"Hey," Mark said. "I was walking Beast and thought I'd stop by." Mark lived a few blocks away and often walked his dog through the neighborhood in the evenings.

I dumped my work bag on the ground heavily. I felt a sudden kick from the baby.

"Great job today," he continued. "Really huge props. It was awesome and everyone's talking about it. I can't believe you did it in two weeks."

That's when it happened.

"Mark . . ." I began slowly. "This was the best day I've ever had at Facebook." I paused for a moment, trying to work out what to say next. And then I was off on a roll.

I blurted out to him that this was what I loved doing. This was how I saw the future of live programming—reaching millions of people with amazing content produced for the web, produced for another generation. That all those days and nights working for this moment had shown me what my true passion was, and that it was something that went beyond my role at Facebook, or even Facebook itself. The entire media landscape was changing—how we got our news, how we experienced live events—and I had to be part of it. I needed to pursue my passion, all the way.

Words tumbled out of my mouth, too fast and unprepared.

"I want to leave."

The words hung in the air. I stopped, suddenly self-conscious and horrified with myself. I had never said any of these things aloud. I didn't even know I truly felt this way until I had spoken the words. I stood there in my entrance hall half wishing I could put the words back in my mouth.

Mark stared at me. Beast stared at me. I resisted the urge to giggle. Beast is too shaggy and adorable to be part of any serious conversation.

If Mark was fazed, it didn't last for more than a heartbeat.

"Are you sure?" he asked calmly, as if he had been expecting it.

I hesitated for a moment. I thought about all the people, all the moments during these past five and a half years on the crazy roller-coaster ride that was Facebook. For a moment, my mind raced through a hundred different memories and emotions.

There was a reason I had said the words I never thought I would say and why I felt so restless. These ideas and feelings had been brewing in me for a long time.

I was going to miss Facebook, but I wasn't afraid to leave. It wasn't the first time I had taken a chance to follow a different path.

The Beginning of the Adventure

I was born in 1982 in Dobbs Ferry, New York. I grew up in a perfectly normal (well, sort of) upper-middle-class family. I am the oldest of four siblings; after me came Mark, Donna, and Arielle, the youngest. Both my parents were doctors: my mom, Karen, a psychiatrist, who had Mark and me while still in medical school and somehow managed the crazy overnight hours of residency while also raising two screaming toddlers, and my dad, Edward, a dentist, whose office was located on the ground floor of our house (how's that for a commute?).

Our town was a quiet suburb about forty minutes north of Manhattan, consisting of 1970s split-level houses laid out along quiet, tree-lined streets, on which minivans carrying soccer-uniformed children drove by. It doesn't get much more typically suburban than Dobbs Ferry, New York.

For most of my childhood and teenage years, right up until college, I led a wonderfully normal life. In the winter, we went skiing. In the summer, my parents shipped the four of us off to summer camp. I begged my parents to take us to "splinter park," a park made entirely of wood, even though every visit consistently resulted in hours of painful tweezer removals. I brought home the chicken pox, and I graciously shared it with the entire family—even Arielle, who was only six months old at the time. And I attended the local Ardsley public schools, dutifully singing the Concord Road theme song every morning, until I switched to Horace Mann School later on.

I guess you could say that I was always a bit of a go-getter. I took piano lessons and somehow negotiated to spend the majority of my lessons singing, while my teacher played the piano. I ran cross-country and had my parents drive our car alongside me on the street so I could feel the pace I needed to run to make the varsity team. I acted and sang in school plays, community plays, summer camp plays—in any show that would have me, really. In high school, I expanded my interests. I became really passionate about studying and singing opera, and I joined the varsity fencing team, where I eventually became captain. I studied really hard in school, got good grades, and in 1999, despite having no special connections or advantages, became the very first member of our family to attend an Ivy League college, when I was accepted to Harvard University. To this day, I remain the only member of our family who has actually graduated.

People always ask me, "How was it growing up with your brother? Could you tell he was going to start a huge company back then?" The answer is a plain and simple no. We were a totally normal, happy family. Besides, don't count the rest of us out yet. I have a feeling we'll all be working for my youngest sister one day.

Fast-forward to April 2003. I had spent the majority of my time at Harvard studying psychology and singing with my beloved a cappella group, the Harvard Opportunes—and now my time at Harvard was ending.

My friends and classmates showed how it should have been done. In those final frenzied weeks after spring recess, every lunch, every party, every dash through Harvard Square was filled with happy, excited people—people with plans, *lots* of plans. And all those lofty ambitions and epic next steps could be summed up with names: McKinsey, Goldman Sachs, JPMorgan, Deloitte. It seemed everyone was heading to Wall Street or K Street to work in banking or consulting. Conversations with friends had become little more than summaries of future résumés.

At some point, the conversation would turn to me. "So, where are you heading, Randi?"

I would smile and look apologetic. "I haven't quite decided yet. I'm thinking of something in the creative industry."

Often these words produced only blank stares. Most of the on-campus recruiting at the time was done by consulting and investment-banking firms, so perhaps it hadn't occurred to my classmates that other types of work existed and that one might even find those other types of work rewarding. (The shock!)

In any case, I wasn't deterred. I had no interest in quantitative analysis or statistics, and the idea of gazing at spreadsheets all day bored me.

Several weeks before graduation, I began hunting for openings at advertising and marketing companies in New York. At one point, my dad excitedly told me that one of his patients worked for J. Walter Thompson advertising agency and was going to recommend me for an interview. It was really sweet; my parents, both doctors, clearly wanted to help me with my desired career path, but they had no direct connections whatsoever to marketing or advertising. Still, they were eager and excited to be able to help in any way possible.

A few weeks later, I walked into J. Walter Thompson and greeted my interviewer with a firm handshake and the necessary amount of confident yet friendly eye contact.

My interviewer beamed at me. "You've come *highly* recommended," he said.

I beamed back. And I was just starting to feel assured of success when I read upside down the Post-it note attached to my résumé lying on the table. "Dentist's kid. Courtesy interview. Thanks!"

I didn't get the job.

Several unanswered résumés later, I landed an interview for the data and statistics team at Ogilvy & Mather, a Manhattan advertising firm. Yes, I know what I just said about statistics. But I did love acting, and I figured maybe I could carry off a convincing impression of an undiscovered prodigy in mathematics.

Needless to say, I completely bombed the interview.

Questioning complete, I began gathering myself to leave. As I headed for the door, a friendly looking man appeared in the doorway. It was the hiring manager for the entry-level program.

"Well, Randi," he said. "Clearly your passion isn't statistics."

I decided to abandon my Oscar-worthy performance and murmured my agreement.

"But," he continued, "you seem creative. A position recently opened up in the client and creative side. Let me just make a few phone calls." A few more interviews, and a few days later, it was official. I had a job.

I had envisioned myself taking a few weeks off between graduating and starting a job, but when Ogilvy called, asking if I could start immediately, I didn't argue. The Monday after graduation, I started working.

At first, I lived at home and commuted to work on the train. It was surprisingly fun and nice being back with my parents and my youngest sister, who was still in high school. But the long commute quickly started to wear on me. Besides, I was eager to start my new life in the city. After a few months of saving up as much money as I could, I decided it was time to move to Manhattan. I was ready for a new and glamorous adventure.

At Ogilvy, I was placed on a fairly new team called "interactive and digital media." Of course, I had imagined myself in the far more exciting surroundings of TV sets and magazine photo shoots. Later, in retrospect, it turned out to be a hugely fortuitous placement. As the power of the Internet grew, my team and my responsibilities grew exponentially, while my friends who had been staffed in the more glamorous jobs were still going on coffee runs. But at the time, I didn't realize how lucky I was.

It was also clear that this was not the creative role I had hoped for. The days were long and mostly spent photocopying, binder filling, three-hole punching, staple removing, and spell-checking memos drafted in legalese.

Even worse, I had a mean boss. She consistently referred to me as "her project" and would do bafflingly cruel things. Once she invited me to give a big presentation to the department head at 2:30 P.M. the following day. I was so excited. Now was my chance to shine! I stayed up late that night practicing my presentation. The next day, she showed up at my desk ten minutes before the meeting.

"Randi, where have you been? The meeting started twenty minutes ago!"

She had moved the meeting and told everyone except me, giving her a chance to reprimand me for being late and irresponsible in front of the department head and bolster her tough-as-nails image. It was the most humiliating experience of my life. (Tip: If you hide in a restroom stall, you can muffle a quiet cry with a courtesy flush and no one's the wiser.)

But I do have her to thank for my Naked Cowboys.

The Naked Cowboys was a tight-knit group of entry-level staffers to which I belonged. We had adopted the cowboy moniker as an homage to the strange, long-haired guitar player who wandered around Times Square every day wearing nothing but a pair of white briefs and cowboy boots. He made a living posing for photos with tourists and raked in a ton of money. He was obviously a secret genius marketer, and he inspired all the shadeball Elmos and nightmare SpongeBobs that haunt Times Square today. It seemed fitting to honor him.

The purpose of the Naked Cowboys was to take on an actual marketing campaign for a nonprofit, above and beyond our daily duties. It was part of a prestigious career-development program within Ogilvy for new(ish) hires, which my boss had recommended me for. It was a win-win for everyone. The nonprofit got a free ad campaign, we got invaluable experience, and Ogilvy got us to work late every single evening for three months, as we were required to work on these campaigns on our "own time." It was grueling finishing a full day of work and then starting another session that lasted until ten or eleven each evening. After we were assigned an ad campaign for the Special Olympics, we worked even later, usually past one in the morning. But the work was fascinating, and I enjoyed being part of a crew. All those late nights and intense sessions together inevitably made us the best of friends.

There were also my other friends. I had taken to life in the city like a duck to water. I took the Craigslist gamble and found a room in a Hell's Kitchen "converted" four-bedroom apartment, which meant I was living in a portion of the living room fenced off by a sheet. So, the apartment was a little sketch. But I loved my roommates, and they became part of my crew. I now had enough people to have fun with. And, boy, did we have fun.

During that time, in the courtyard outside Ogilvy & Mather, I shared a bunch of margaritas with a wonderful guy (and Harvard classmate) on an amazing first date. I didn't know it yet, but he was the man I would end up marrying: Brent Tworetzky.

It was 2003, and I was twenty-two years old. The city was hot and vibrant. I was living paycheck to paycheck with no savings to speak of. But I was young, surrounded by friends, and making my own way in New York. I was happy.

•   •   •

Against this backdrop, I was dimly aware that Mark had started working on a new project: Facebook. Or as it was known back then, The Facebook. The site had gone gangbusters at Harvard and had started to spread to several other U.S. colleges.

Occasionally, I would ask my colleagues whether they had heard about The Facebook. Everyone my age was familiar with it, but no one over twenty-four seemed to have a clue.

Still, Mark was clearly on a roll with the site. He had just flown out to California to secure funding for his company. He asked me if I wanted to come work with him.

Despite my misgivings about corporate life, I politely declined. I hadn't given up on New York just yet. But as time passed, I continued pondering my future plans.

So, that was my life as a New Yorker. I assumed the experience was uniquely mine, that all the joys, ambitions, worries, and dilemmas were special to me. But of course they weren't. All these things would be instantly familiar to young professionals who have ever dared to strike out on their own in the big city. The experience of being a newcomer in New York, despite growing up so close to the city, is one shared by tens of thousands of new grads, interns, and struggling-artists-cum-hipsters every year. I lived a life that was no different, no more exceptional. But that's the thing about New York. The unexceptional thing about most new arrivals is that they all want to be exceptional—and that was certainly true for me.

I didn't have a lot of money or connections. I didn't have the dream job, the dream apartment, or the dream career mapped out. But I still had a dream, as vague and ill-formed as it was.

I knew that I wanted to do something meaningful with my life. I wanted to do something that would have a big impact—on the kind of scale to touch a lot of people's lives. I wanted to do something that would let me reach my full potential. And I was still convinced that I wanted to do something creative and connected with the arts and entertainment.

A few months later, I left Ogilvy. I was ready to move on, but I didn't join Facebook just yet.

Instead, I took a job at Forbes. I was offered a role producing a really intense TV show called *Forbes on Fox,* during which four older guys would yell at one another about the economy for sixty minutes every Saturday at five in the morning. Sitting in the control booth, I'd worry about one of them having a heart attack and do my best to ignore the shouting. Still, I was excited for the chance to work with Steve Forbes and figured that it would be an interesting experience.

At the time, it felt random, but little did I know that this brief foray into television and video production would open so many doors for me and shape and define so much of my career to come.

After my first show, Steve Forbes offered to take me out to breakfast. I tried to contain my excitement as we left the Midtown studio. *Where did übermillionaire Steve Forbes eat breakfast?* As it turns out, I didn't have to contain my excitement for long. He took me to the nearest restaurant—a Wendy's inside the Forty-Seventh Street subway station—which I quickly learned was his go-to spot, post show filming.

"Order anything you like!" He beamed. "Even biggie size."

During an hour-long discussion about the Yankees, I remember thinking how cool it was that such a successful, accomplished businessman had remained so down to earth.

A few months passed. My brother reached out to me again to join him at Facebook, now operating out of a house in Menlo Park. *Interesting*, I thought. *It's no longer The Facebook. He's making progress.*

"Why don't you come out and see what we're building?" he asked. I hesitated. "Maybe," I said. "Let me see if I can get a cheap ticket." Mark was decisive. "I'll get you the ticket. Just come."

And so I did.

First Days in California

A couple of days after Mark's call, I flew to San Francisco for the weekend. As promised, he had taken care of my ticket, booking me on a JetBlue flight that landed late in the evening. A driver waited for me in the arrivals lounge at SFO, holding a sign with my name on it. And then I was whisked away in a black sedan, down to Menlo Park. I was awed by the novelty. I had never taken a car service before.

And there I was, half an hour later, on a dark, remote street, thousands of miles from home, in front of a nice, unassuming suburban house. It really didn't look like the secret headquarters of a future technology empire.

I walked up the path and knocked on the door.

Mark answered. "Hey! You made it! Come meet the guys."

We walked into the next room. You could tell this was a really nice house, perfect for a family and just begging to be filled with tasteful fittings and good furniture. But all those thoughts vanished at the sight of the living room.

The lights were off and the room was dark. The only light came from the half dozen computer monitors clustered around a crowded table in the center of the room. And there, working off a tabletop overflowing with empty drink cans and plates of half-eaten takeout, were four very disheveled guys.

"Guys, this is Randi," said Mark.

There were grunts of acknowledgment and a couple of waves. All of them were intently focused on their screens and wired in with their headphones. Introductions had to wait until a little later when we went out for dinner at the Dutch Goose, a local dive bar.

I remember being slightly shocked at first by the general grunginess of the lifestyle of Mark and his team. The house and the bar summed it up perfectly. The food at the Dutch Goose could be charitably described as semi-edible, but the vibe was energetic and fun, and the beer flowed freely.

That night, Mark started laying out the full scope of everything the team was working on. And I remember just how serious and passionate he was.

"We're going to connect everybody," he said.

Over the next few hours he talked more about the features they were building and their plans for campus expansion. Mark has always talked fast; it helped catch me up on the details quickly. The rest of the team were mostly content to listen in, knock back beers, and talk among themselves about the current items they were coding.

Mark completely blew my mind, and the awesomeness of the Facebook vision was immediately clear to me. Over the next couple of days, I found myself viewing that grungy lifestyle with a mixture of awe and respect. A group of guys all living in a house together coding around the clock didn't sound like fun to me at first. But these guys had such an incredible belief in their work and an amazing laser-like focus, I couldn't help but be reeled in by it and by them. I realized that I was literally watching the American dream play out right in front of my very eyes, one Red Bull at a time.

They *were* going to connect everybody, I thought. And then I sat back and watched them code. It was insane. It was brilliant.

But my true aha moment happened when I was invited to join a meeting about some key Facebook marketing materials. The point of the meeting was to solidify some of the visual design features for Facebook—the general look and feel and color palette. This was amazing. I was a fly on the wall in a meeting to decide how a network for five million would look. A debate was going on. I leaned in to pay attention.

And then suddenly everyone was staring at me. "Hey, Randi, you're the marketing person. What do you think?"

Ten years of a career I hadn't yet had flashed before my eyes. It would honestly have taken me a decade at Ogilvy to even be invited into a room where a conversation like this was taking place, let alone be given the chance to be a decision maker.

I cleared my throat. "Well, here's what I think."

No one interrupted or laughed at my opinion. After I had finished laying out my views on my preferred shade of blue, along with a few other marketing ideas, the debate resumed—with me included.

I can't even remember how that debate concluded. My overwhelming memory is that my heart was soaring. In that moment, I realized just how amazing Facebook was as a career opportunity. I knew I had to take it.

On the final evening before I flew back to New York, instead of spending it at dinner, drinking, or trying to be useful while the guys coded, I spent it sitting in Facebook's new official office, located above a Chinese restaurant in downtown Palo Alto, negotiating my starting salary with my brother. We sat across from each other at his desk while he decided a salary and stock-option grant for me on a napkin.

"How about this?" He slid the napkin across the table.

The stock was good. But why go for equity over real money? I crossed out the stock options and bumped up the salary. I passed back the napkin.

Mark gazed at it for a moment, then made a decisive gesture across the paper. He scribbled for a moment and handed it back.

He had rejected my numbers and restored his original offer.

"Trust me," he said. "You don't want what you think you want."

I didn't recognize it at the time—I was twenty-two years old and all I saw was a chance to make more than the biweekly check of $900 I currently earned—but I sure as heck recognize it now.

Years later, I would stand in the entrance hall of my home, pouring my heart out to my brother about how I was ready to leave Facebook. But on that fateful summer evening in 2005, in the quiet calm of the empty Facebook office, a new chapter of my life was about to begin.

Today, people often ask me, "Now that you know what you know, what would you go back and change?" It's a silly question. I don't know if they expect me to impart some holy wisdom or if they expect me to admit to some grave screw-up I made along the way. Usually, I just crack a joke and say, "I would have asked for more stock." It always gets a laugh from an audience, but every time I say that, I think about that evening negotiating with my brother, my *younger* brother, and how he looked out for me, even though I was too young and naïve to recognize it at the time.

With the important contract details safely concluded on the napkin, it was time to start imagining my new life. I grinned the whole flight back to New York.

"You're looking happy," said the old lady sitting next to me in my economy seat.

I beamed at her with such intensity I think I slightly scared her.

I was so excited about this new life direction and career path that

I deluded myself into thinking that everyone back in New York City would be over-the-moon thrilled for me, that balloons and cheering parties would await me at LaGuardia Airport.

Well, I was in for a bit of a brutal awakening. My teammates at work told me I was making a horrible mistake and throwing my career away. Brent, who I was now dating, had just given up his dream job in San Francisco to be with me in New York. He wasn't exactly pleased. Over a long, tearful dinner at Mr. K's Chinese restaurant in Manhattan, we discussed what this could mean for our relationship.

My mom was excited for me. It's secretly every mom's dream that her kids work together. She encouraged me to take this terrific career opportunity and supported my pursuing my dreams. But she also pointed out that I had a good thing going with Brent and told me to put real effort into a long-distance relationship. She had taken a real shine to Brent. After the first time she met him, she had called me to say, "Randi, do not f*** this one up."

Now, as I prepared to leave for a new life in California, and at Facebook, my mother called to give me one last bit of friendly career advice.

"Randi, good luck in California. Do not f*** this one up."

chapter 2

THE DIAMOND IN THE ROUGH

And so I came to California and to Facebook.

I officially joined on September 1, 2005. But it was a few more weeks before I was in the office full-time. After hurriedly wrapping things up in New York, I still had to transfer my life to the West Coast.

Soon enough, I found a Craigslist posting for a room in a house in Menlo Park with three grad students, which didn't sound *too* shady. The location was perfect, and that was enough for me. Without seeing it in person, I took the room.

If I had been starting any old job at a more conventional company, the experience of uprooting my life and adapting to a new life in suburban California might have been more daunting. Besides Mark, I barely knew anyone in the SF Bay Area. At Facebook, I didn't have a team and was one of only a few non-engineers. Our offices were pretty modest—the usual start-up digs—and located above that sketchy but surprisingly good Chinese restaurant and a deep-dish pizzeria.

But I never felt lonely or bored or out of place. I arrived at an amazing time. Shortly after I joined, we celebrated reaching five million users. Investor Peter Thiel threw us a party at the Slanted Door, a fancy Vietnamese restaurant in San Francisco. I remember sitting at the table in that small, happy group, feeling we were on top of the world. Five million users! How could it get better than that?

Even if I had wanted to stress, I didn't have the time. From the moment I stepped through the door on my first day, I was rushed off my feet. The company was only a few dozen people, so the arrival of one new person was a substantial increase in capacity. Everyone wanted a little bit of help. Just a small favor. Pretty please?

I was not opposed to being bribed with cupcakes.

I pitched in wherever needed. I relished the chance to experiment with my role and live the start-up experience. In the early days of a start-up, it's not uncommon to hold many jobs and wear many hats. Toward the end of my time at Facebook, I joked that I had worked on every single team except the IT help team. For a few months, my business cards listed "samurai warrior" and "ninja" as my title, because I was working on so many different teams it would have been confusing to simultaneously list them all on one card.

As a non-engineer, I was given non-engineer stuff to do. In the early days, I had a hybrid marketing–business development–sales role. We weren't doing much "traditional" marketing at that point, so I was supporting several other teams within the company. Facebook was still only available at American colleges and universities, and the site was succeeding so well within this market that it was doing a pretty excellent job of marketing itself. In fact, my total marketing budget for the first year at Facebook was about a hundred dollars, which I used to print up T-shirts for students when we filmed a video on the NYU campus. And I probably overpaid.

Most of my first year was spent working with the sales team, helping build larger marketing packages and campaigns that they could sell to companies that wanted to reach students on Facebook. I helped put together our first back-to-school campaign, building a back-to-school hub with companion ads. It was the biggest sales and marketing program we had done to date.

Because the company was small, we worked all the time. Days bled into nights and into weekends. Hours at the office turned into hours hanging out at someone's house or The Old Pro, our favorite bar in downtown Palo Alto. Colleagues became close friends, and several "Facebook couples" arose.

The early team felt an intense closeness—the kind of bond you form when everyone works together toward one unified goal. I knew that it would probably be a fleeting experience and that one day we would all go our separate ways, but it was awesome at the time. Facebook was our job, our community, our social lives—our lives, period. And I loved it.

We shared pride in the mission and proudly wore our finest Facebook swag—hats, T-shirts, hoodies. To this day, no designer handbag, bought during a moment of shopaholic weakness, has ever elicited as many comments from strangers as my old Jack Spade Facebook laptop bag.

In New York, it seemed everyone kept an almost religious separation between his or her career and social life. In California, it felt like the company was one huge family.

It wasn't all roses, though. During that first year in the company, I faced two major challenges. First, I had to get used to the fact that while marketers and business folks ruled the roost in New York, in Silicon Valley they were the background noise. Out west, engineers are thought of as rock-star gods. Everyone else is barely a roadie. If you weren't the one doing the coding, you had to shout pretty loudly to be heard.

Second, there was the issue of my being Mark's sister. I worked my butt off, and I bled Facebook blue, but no matter how hard I worked, many thought of me as just the boss's sister and assumed I was there only because of nepotism. One colleague even referred to me as "Zuck's sis" for almost a year. I acquired a lot of new best friends as soon as they were hired and lost them as soon as they realized I didn't have Mark's ear on anything product related.

And I didn't have it any easier from Mark, either. The first meeting I attended with him, one week into my job, he tore up a piece of paper that I presented to him, in front of everyone. I remember him walking past my desk in the open-plan office a few weeks later, saying hi to everyone except me. When I asked him about it later, he had this response: "You know, I never really thought about it. But I guess I sort of feel like I have to go out of my way to *not* be friendly with you, to show people you're not getting special treatment."

Thanks, bro.

One of my colleagues, a few years later, summed it up well. "You know, Randi," she said. "I'm sure being Mark's sister opens a lot of doors for you, but I don't envy your situation. Women in technology already have to be twice as good as their male colleagues to get to the same place. But you have to be three times as good. And *even then*, people are still going to question your success."

The benefits of my last name far outweigh the negatives, and it's definitely opened lots of doors for me. But open doors alone don't get you anywhere unless you do something once you walk through them.

From the moment I joined Facebook I knew I had a long road ahead of me and was existing beneath a very big shadow. There was a chance that no matter what I did I would never be more than someone's sister.

I was twenty-three years old when I joined Facebook. As a Harvard grad living in New York, I *felt* like I knew it all. But I was still a baby, and there was clearly a lot I didn't know about navigating the politics of a new workplace.

I got my first taste of the limelight when a well-known tech blogger, Robert Scoble, wrote an article about me called "The First Sister of Facebook." I had reveled in the spotlight my whole childhood, doing theater, music, and singing a cappella, and I found this attention addictive. I celebrated my newfound status in the Valley, created parody music videos and posted them on YouTube, and developed an exaggerated reputation for enjoying a cocktail (or three) and grabbing the microphone to sing at pretty much any event that would allow it.

My go-to song at Facebook company parties was "Bring Me to Life" by Evanescence, in which I would duet with my close friend Chris Kelly. One of my finest contributions to Facebook company culture was when our employee cover band, Evanescence Essence, for which I was the lead singer, won the first ever in-office version of *American Idol,* "Facebook Idol." Our motto was "Evanescence has two hit songs, and we do *both* of them."

Because I was young and new to Silicon Valley, and I had never experienced this type of company culture before—where your coworkers are your friends, family, and community—I made the mistake of treating my colleagues more like college dorm mates. It was an easy mistake considering how many waking hours we spent together and how close we felt, but I started to believe that I could just let my hair down and be my true outside-of-work self way too early on. In reality, though, I was still making my first professional impression on everyone and should have held my cards a bit closer to my chest.

Maybe you can get away with being a fun-loving person as well as a respected professional if you're a guy, but I'm not sure that's true for a young woman. One of my most respected mentors took me aside and said, "You know what, Randi. Because you're a woman, they're only going to talk about you in one light in the press. Do you want to be the brains behind Facebook's marketing strategy? Or do you want to be Mark Zuckerberg's ridiculous sister who sings?" The honest answer is that I wanted to be both. I wanted to live in a world where you could be a successful executive *and* have hobbies and interests that made you more than a two-dimensional person.

If I had to do it all over again, I'd have kept my head down and focused on work those first few years and let people get to know the work I was capable of before they got to know my "creative side." Knowing what I know now, I basically did everything that I would advise a young woman going into the workforce *not* to do.

In spite of my flamboyant behavior, I had some loyal champions inside the company. Thanks to them, I eventually found myself in a series of roles that played to my interests, my passions, and my creativity, plus enabled me to leverage my outside-the-box thinking and my love of the limelight to best benefit Facebook.

At first, Mike Murphy, the affable head of the sales team, took me under his wing and made it his mission to give me projects that I could own and shine with and use to shift perceptions of myself within the company. One of my favorite early memories at Facebook was planning the back-to-school campaign with Mike on the back of a napkin. (Students could assign fun superlatives, such as "most likely to succeed" or "most likely to live in Kansas," to their Facebook friends, while also viewing great deals from some of our very first advertisers.)

At start-ups, a lot of work gets accomplished on napkins.

Then, in mid-2006 Facebook brought in a head of business development, Dan Rose. One of the first major deals Dan signed was a multimillion-dollar media partnership with Comcast, Facebook's largest to date.

One day not long after he had arrived, Dan came to my desk. "I hear you're pretty creative and like to work with the media," he said. "Want to come work with me?"

Soon after, I joined the business development team to help negotiate and manage the Comcast deal. I worked with Dan for about two years, leading deals with media partners from Comcast to ABC News to CNN.

Over those two years, while the rest of the business development team was doing mergers and acquisition deals, striking partnerships and negotiating an investment from Microsoft, I specialized in developing partnerships for Facebook with mainstream media and television outlets. And when in late 2008 it was finally time for Facebook to have an official marketing team, I joined forces with two amazing women, Raquel DiSabatino and Meenal Balar, to create a brand-new group called consumer marketing, which is where I remained until I left Facebook.

Being asked to help create a marketing team from scratch at the current hottest tech company in the world was an incredible and humbling moment. Three years earlier I had been restless and frustrated in New York, dreaming of a day when I would get to really test myself and show what I could do. I imagined a time when I would get to lead my own team and develop my own plans—and then drive the change that I wanted to see.

In the summer of 2005, I had walked into a room of half a dozen engineers hunched over their computers. The empire that was Facebook in those days extended from the kitchen to the sofas in a single suburban house. But in the years since, those few guys had become several hundred employees, drawn from the best and brightest talent across the industry.

Pundits, the media, academics, celebrities, and the entire World Wide Web were constantly debating, criticizing, and pondering the future of the company and our impact on the world. In 2010, journalist and author David Kirkpatrick coined the phrase "the Facebook effect" to describe the unique role Facebook played in igniting global attention and support for causes and content. It was a concept that captured the zeitgeist perfectly. We were the future of friendships, dating, business, marketing, entrepreneurship, activism, philanthropy, and revolution.

Technology Is Not the End—It's the Means

In Silicon Valley, it's easy to end up with a narrow, tunnel-vision view of the work you're doing and an insular perception of what technology means. It's easy to get so focused on site data, product reviews, and industry blogs that even the smartest people on the planet, building some of the most world-changing products in all of history, often forget that there are actual human beings on the other side of the technology.

The Valley—by which I mean the entire tech industry in the Bay Area, including San Francisco—is at the forefront of all the major innovations currently taking place in connectivity. It's where so many major technology companies are located, including Apple, PayPal, eBay, Google, Facebook, LinkedIn, Pinterest, Twitter, and Yahoo! The result is a robust ecosystem for start-ups, investors, and tech bloggers. A place where fearsome rivalry for top engineering talent is constantly brewing, and companies compete to offer crazier and crazier perks to keep their current employees satisfied.

And where the term "golden handcuffs" refers to people who would prefer to work elsewhere or start their own companies but are making so much money they often decide to just stay where they are. There's no separation between work and play. If you live and work in Silicon Valley, you eat, sleep, and breathe tech 24/7. You read about it in the news, hear about it in the gossip at the coffee shop, and feel it in your friendships and relationships.

As I developed in my career at Facebook and came to better understand the Valley, I realized that it wasn't the coding or the systems or the tools that got me jumping out of bed each morning. And it certainly wasn't the corporate rivalry and endless battles to be number one. I cared about the people on the other side of the technology. I cared about what we could *do* with the tools and what the industry could do together. And, more than anything, I cared about the impact we could have on people's lives, in communities around the world.

I knew that it was going to be an uphill battle to push people to see the humanity behind the tech, especially in a company and industry where engineers and data reign supreme. It was an uphill battle I was willing to fight. I could see Facebook's incredible ability to strengthen friendships and connections between people. I was convinced that we had an opportunity to create value in even more ways, from education to art, and from science to business and community. The day President Obama came to Facebook epitomized all that I knew to be true about the power of connectivity.

But the moment when all these ideas first began to come together was earlier, in 2007—when I saw how Facebook could be used more for politics than Poking.

It was the summer of 2007, and the frenzy of the next year's general election was about to kick off. It would be the first presidential election since Facebook and social media had exploded. And even though it was many months before their campaigns would begin in earnest, many of the would-be candidates had reached out to Facebook to help them set up public pages, which would go live when they announced their candidacies.

I suddenly found myself privy to all sorts of fascinating information. Because of all these candidate inquiries, my team and I had a good sense of who was planning to run for office, even before many major news outlets did. That summer, when pundits were speculating about whether Mike Bloomberg was going to run for president, I enjoyed the wry satisfaction of knowing that his team had not set up a public Facebook page yet, and even though I didn't know for sure, I could guess that he was therefore not likely to be entering the race.

But there was something else I realized. Facebook was evolving from being a site just for college students to being an indispensable tool for everyone, including those who wanted to do something really important, such as influence the outcome of a presidential election. This was also an opportunity for Facebook to expand its audience and influence.

I knew we had to get more involved in politics, beyond just helping the candidates set up public profiles. When I began preaching internally about the importance of Facebook getting involved in an event that would dominate the next year and a half of mainstream media, I assumed that everyone would get the idea pretty quickly. A fair number of employees were excited at the prospect. But that summer the company was developing a complex new advertising system, set to launch before the end of the year. Advertising was the shiny, sexy project to be working on, because it had the undivided attention of Mark and the rest of the management team. This meant working on the ad product was a chance to impress the bosses and *maybe* even get face time with Mark, which was the holy grail of Silicon Valley geekdom.

My colleagues quickly lost interest in an election that was still over a year away.

At this point, you may think that the smart career move would have been to refocus, shift my priorities to align with everyone else's, perhaps even play some "corporate politics." And you'd probably be right.

Which means . . . I set off to work on politics on my own. For better or worse, I've always gravitated toward finding my own diamond-in-the-rough project that I can own and shape, rather than what's shiny, new, and exciting to everyone else.

Luckily, I had the support of four friends and colleagues: Dan Rose, my old boss; Ezra Callahan, a fellow early Facebooker and influential product manager; Chris Kelly, our general counsel; and David Fisch, a leader on the business development team. They became my full and willing collaborators in this crazy scheme. Later on, Adam Conner and Andrew Noyes joined Facebook in Washington, DC, and became my partners in crime and lifelong friends.

Our first move was to team up with ABC News.

Our opening discussions took place as competing networks and tech companies began maneuvering to achieve their own vision of the first social media election. YouTube and CNN were the first to announce they were producing a debate together, which immediately launched a social networking arms race between the other networks. The news media were desperate to have their own Internet tie-ins for their shows and to find some novel, interactive element to engage younger, tech-savvy audiences.

It quickly became the Wild West. Grab your favorite social networking site and go! Television networks were announcing "integrations" with every possible Internet site under the sun, desperately hoping something would stick. We took a lot of calls from companies pitching half-baked partnerships, which we universally declined. We wanted to do something smart and unique that would really show off the value of Facebook's platform.

And then we found someone who saw eye to eye with us. Over a three-hour lunch at Buca di Beppo in downtown Palo Alto with Andrew Morse and Paul Slavin of ABC News, as we consumed an inhuman amount of Diet Coke and a platter of meatballs larger than my head, we came up with the idea of holding a joint online-offline presidential debate before the primaries, which would be broadcast on ABC News and feature Facebook updates and polling results on-air. Additionally, we would develop a social politics app on Facebook that would allow users to discuss and debate election issues with one another, and ABC News would feature answers from the app during the debate.

This direction appealed to me because it was actually useful and it offered real and unique value. For Facebook, it was the perfect showcase for both our "News Feed" feature and our newly launched platform for applications. Whereas most early apps were games, this app would allow users to post their views on the most important topics of the 2008 election and see where their friends stood. And whenever they posted a viewpoint, it would show up in News Feed, so their friends would see it and be able to respond and engage in debates. For ABC News, it was a way to be young, fresh, and different from the competition. We quickly moved to sign a deal.

That part of the process, however, was a little more stressful— and memorable—than we had expected. I was working closely on the contract negotiations with David, my colleague on the business development team, and Julia Popowitz, on the legal team. Facebook had moved into a brand-new office in downtown Palo Alto. The electricity hadn't even been turned on in the building yet, so we spent countless evenings using the light from our BlackBerry devices to pore over the dense, thirty-page contracts sent over from ABC. Negotiations also got so heated at one point that someone even had a brief heart attack scare.

Eventually, with the horror of negotiations over and the papers signed, the core team got to work. On the Facebook side, my colleague Ezra and I, along with a handful of engineers, focused on building the app. On the ABC News side of the house, we were introduced to two guys who would become our cohorts over the next year: Austin Vance and Bradley Lautenbach. They had just wrapped production on a show about YouTube and therefore were the de facto cool, hip, tech guys inside ABC News. This pairing was highly serendipitous, especially when you consider that five years later Bradley and I would go on to start Zuckerberg Media together. But sitting in a Houston's restaurant in New York City, guzzling Arnold Palmers on a hot summer afternoon, all we could think about was how to disrupt politics by bringing "old media" and "new media" together.

Fast-forward to January 2008. Our U.S. Politics app had grown to a million users, which was a really big deal in the early days of the Facebook app platform. Hillary Clinton went head to head with Barack Obama in a debate held in between the Iowa caucuses and the New Hampshire primary. And right there, on-air, Diane Sawyer pulled stats and answers live from Facebook users. The team was exceptionally proud, and even more excited that the moderator podium was emblazoned with a giant Facebook logo.

There was something else I was excited about. The day before the debate, I did my first live TV appearance—on *Good Morning*

America, no less. I was in New York with Ezra when the ABC team gave us the chance to attend a live taping of the show. I had never been on the set of a national morning show before (not counting the 5:00 A.M. Steve Forbes show on Fox). Ezra and I woke up at the crack of dawn and headed over to the studio in Times Square, where we hung around on the set gawking at everything. As we were there to just get a backstage tour, and it was an ungodly hour of the morning, we were wearing our "Silicon Valley finest," by which I mean Ezra was wearing a Facebook hoodie and jeans and I was only slightly more dressed up in a casual sweaterdress and slouchy boots.

The producer came up to us. "So, we have this segment on the show where we slot in local weather in a few places across the country, but anyone who doesn't get the weather just sees the anchors bantering for a minute. Why don't you and Ezra do the banter with us for that minute?"

Our mouths fell open.

It was three in the morning back in California. Facebook's PR team was sound asleep, and even if they had been awake, it would have been impossible to get their blessing in time for the appearance. We needed to make a split-second decision. I glanced at Ezra. He gave me a knowing look, and we said yes. When would we get a chance to be on *Good Morning America* again? We went ahead and filmed our minute of banter.

Everything went well. Afterward, I took some heat from the PR team. But hey, sometimes it's better to ask forgiveness than permission, right?

My confidence that we were really on to something only grew when Mark Penn, Hillary Clinton's pollster, publicly declared that Obama's supporters "look like Facebook." He didn't mean that as a compliment, but that Facebook was something young, inexperienced, and irrelevant. Shortly after, Obama won Iowa, with his surprise success largely attributed to young voters energized by social media. Now Facebook wasn't just playing a part in shaping the presentation of the campaigns; it was influencing the campaigns.

My diamond-in-the-rough project was starting to look pretty shiny.

After the debate, we started hunting for something new. Facebook didn't have the resources to keep the U.S. Politics app going, and ABC didn't want to take it over from us. The app ended up being shut down and the code deleted, which was hard for me to swallow. It seemed a waste of a million followers and a lot of hard work. I was keen to develop something that was built to last.

That's where Andy Mitchell at CNN came in. Andy and I had begun talking several months earlier when I had accepted an invitation from him to fly to St. Petersburg, Florida, to attend the second CNN-YouTube debate as his VIP guest. At the time, Andy had been trying to woo Facebook away from ABC News. Now I decided to try to strike up the conversation again.

Luckily, Andy was more than receptive, and we tossed around countless ideas with CNN. Should we do another debate? Should we do something with polling and stats? In the end, we opted to extend the social politics app into a full microsite on CNN, which people could log in to with our newly launched Facebook Connect button. The site would feature live debates between Facebook users on election night, and it would take what we had done with ABC to the next level. Even though people would be posting their thoughts on CNN .com, they would show up on Facebook, encouraging their friends to join them. This would be a showcase for Facebook Connect and a win-win for Facebook and CNN. How could it not be?

The folks at CNN were so excited they started blasting out television commercials to promote the microsite. Anderson Cooper even recorded a promotional video. This was my very first project under a new boss, and I desperately wanted it to be successful. Andy and I were both laying a lot of our professional capital on the line with this project.

Unfortunately, it was a big disappointment. There were multiple technical problems on both sides. On election night, people seemed to be on every other site *except* ours. You could almost hear the crickets chirping in the forums. The microsite was difficult to find on CNN's website, and the Facebook Connect button did not provide a very compelling experience if nobody was there to connect. We had thrown away a huge opportunity to define the coverage of a momentous day.

But then from the jaws of defeat came victory.

That same week we had a "hackathon" at Facebook, an event that happened every few months, when people stayed up all night, working on passion projects, and then presented their projects to the entire company the next morning, followed by a pancake breakfast. I know it doesn't sound fun to voluntarily pull an all-nighter at work, but these events were energizing, exciting, and community forming, and they embodied the passion and entrepreneurial spirit of the company.

During this particular hackathon, it was about two in the morning and I was getting ready to head home (as hard as I tried, I was never really one for all-nighters) when two engineers came up to me.

"Randi, we know you're working with the TV networks. We have an idea."

The engineers, Peter Deng and Ari Steinberg, explained that during the debates, they found it annoying that they had to keep two browser windows open on their computers at the same time—one to watch the debate and one to see what their friends were saying about the debate on Facebook. Why didn't we build something to combine them?

I called Andy Mitchell.

Four weeks later, we had a working prototype of the system, and CNN planned to use it for their first socially enabled coverage of the presidential inauguration of Barack Obama. But it wasn't an easy journey. It took a lot of resources from both CNN and Facebook. There were huge technical challenges to solve. We had to blacklist over ten thousand potential spellings of every conceivable curse and swear word (this was going to be a family-friendly Facebook inauguration, gosh darn it!), and there were questions about whether the site would be able to support so many millions of concurrent video streams at once.

In true Facebook–Silicon Valley spirit, a random late-night conversation with two engineers had turned into a real product in record time.

But there was more work to be done.

A week away from the inauguration, I was on my daily phone call with Andy, discussing how the app would be used on the day, on-air. CNN wanted to fly one of their anchors out to California to cover the inauguration from Facebook HQ, a Silicon Valley political correspondent who could discuss the topics buzzing on social media and the online reaction to the big day. They were having some trouble finding the right person, though—someone who "got" social media, was young enough to appeal to a youth audience, was old enough to be taken seriously, and was approachable on camera.

I made some sympathetic noises down the phone. Someone like that would be tough to find.

"Want to do it?" Andy asked casually.

I paused. "What?"

I could literally feel my heart pounding. The limelight-loving theater geek in me wanted to scream, "*Yes!* Oh God, yes!" but I was scared that my colleagues would think I purposefully orchestrated this to get attention. I asked Andy to give me a bit of time to think about it and run it by a few people. He agreed but asked if he could still float my name past his bosses to get their reaction, as time was short.

I paced around the office while I considered my options. I thought about being back in New York, during those early days after college. I thought about the long days and nights at Ogilvy. I thought about that first trip to visit Mark in California and seeing that handful of engineers coding in the dark living room. I thought about all the silly videos I'd made and how much I loved being in front of the camera. And then I thought about all the different projects I'd worked on in the past four years at Facebook, up until this election platform.

It was insane that it had come to this: me, on television, talking about technology and politics in front of a live audience of millions. It's one thing to spend a minute laughing and bantering as a program filler. It's another to play the pundit. I thought I had given up on my TV/theater dreams when I chose to pursue marketing. And now here I was, with the chance of a lifetime being dropped right into my lap. The amount that I wanted this to work out terrified me, because it forced me to come face-to-face with a part of my identity I realized I would never be able to fully silence. I loved the idea of being in *front* of the cameras.

I was terrified of what allowing myself to love that would do, or mean, for my career. But I knew I'd never be happy just being behind the scenes, just being someone's sister.

Andy called back. "The producers love the idea of you doing

the show. Are you in?" I had received the go-ahead from Brandee Barker, Facebook's PR director, and it was time to seize the incredible career opportunity before me.

"I'm in."

The evening before, I had had a flashback of a childhood memory. I was in fourth grade and it was during Operation Desert Storm. Our classroom was covered in yellow ribbons and you could hear CNN playing from every television set in our neighborhood. One night during dinner, my mom asked me if I knew what was happening on the news. I said no. She said that she was disappointed in me and that staying up with current events was a very important life skill. I was so desperate to change her opinion of my being uninformed that I didn't leave the living room for the next three days, as I watched CNN and took copious notes.

As this memory came rushing back, I picked up the phone and dialed my mom in Westchester, New York. "See, Mom. I am up on current events. I *am* the current events," I said.

"Huh?" She clearly had no idea what I was talking about, almost twenty years later, but it didn't matter. She wished me good luck and I attempted to get some sleep, knowing that with all the nerves and excitement, sleep was a distant possibility.

On Inauguration Day, I arrived at the Facebook office at 2:30 A.M. I was there for my first on-air segment of the day, at 3:00 A.M. Pacific time/6:00 A.M. Eastern time. Wearing a pin-striped suit with a silk blouse, I was decidedly *not* Silicon Valley. But luckily, the scene at the office felt familiar and energizing. Even at that ungodly, antisocial hour, the place was humming with activity. Many engineers were still there from the evening before.

Laura Barnes, a Facebook executive assistant who used to work for MAC Cosmetics, was cheerily waiting to do my makeup. Jeff

Rothschild, a senior executive on the engineering team, had set up a "war room" with a technical team who would ensure that Facebook remained up and running, even while millions of people simultaneously streamed hours of live video. Engineers Tom Whitnah and Luke Shepherd were troubleshooting with CNN's team. Tim Kendall, from the ads team, and the first person I would be interviewing on-air, was going over a few talking points for our segment. And there were all sorts of other people from the Facebook and CNN teams scurrying around and making the broadcast work.

At 3:00 A.M. on the dot, they counted me down.

I can't remember what I said during that first segment. I have a feeling it was a total train wreck. I didn't know how to hold the microphone, and I was distracted by people talking into my earpiece while I was supposed to be delivering the news. Apparently, this whole "national correspondent" thing was way harder than it looked. I assumed they were going to cancel the rest of my segments.

They didn't. I kept going. And as the day went on, I grew more confident. When they asked me to talk on the fly for longer, I was able to conjure up interesting filler remarks. I bantered with all the big-name correspondents, and I liked it. And even when my earpiece malfunctioned at one point, I still managed to roll with it. I even announced some breaking news, when Senator Ted Kennedy collapsed during the inauguration. I broke the news on Facebook before many of the other anchors had it.

When the day ended, I was exhilarated and euphoric. I hadn't screwed up! Best of all, we had achieved twenty-six million concurrent video streams through our system, and CNN had captured four times as many viewers as the other networks, which they attributed to their integration with Facebook.

Later that week, at Facebook's monthly all-staff meeting, my team was asked to stand by Chris Cox, VP of product, and the entire company gave us a standing ovation.

That's the thing about working on the unsexy project. The sexy projects usually have tons of cooks in the kitchen, lots of people standing ready to take credit. But when an unsexy project goes well, it's usually pretty obvious who's responsible for the success. My tremendous investment had paid off. I can still remember exactly how it felt standing up in front of the entire company as Chris said, "This project was Facebook at its best. It was a win for Facebook, a win for CNN, and a win for President Obama." After a year and a half of hard work, my idea had become the shiny, new thing.

When this election cycle had begun, almost all of the pundits were questioning whether social media had a meaningful role to play in politics, either in the way news was reported or in the way voters participated in the campaigns. Up until then, they had a point. Many of the social media and online tie-ins that broadcasters had tried previously had been novelties, failures, or both. Even our own Election Day microsite had fallen into that category. But we had shown convincingly that social media provides real value for broadcasters and viewers. We had demonstrated the power of combining on-air and online during the debates. And we facilitated the first truly social presidential inauguration.

We had done more than just provide a good model for the media industry. We had shown how innovation and politics could go hand in hand. We had proved that technology, when combined with broadcast television, could become a potent and entirely new force for engaging and mobilizing voters. After the election, as analysts began to examine why Obama's campaign had been so phenomenally successful, particularly with energizing the youth vote, social media and the Internet were largely credited for his success. And Facebook's efforts with the U.S. Politics app, the debates, our politician profiles, and more, were always mentioned in those analyses.

Through this foray into the unknown, I had found a new career path—part product manager, part producer, part hacker, and part television correspondent. And from that day in January 2009, my life began to quickly move in a new direction.

This Book Isn't for the Techies—It's for You

And so we return to the beginning of my end at Facebook, in the entrance hall of my house in April 2011 with Mark and his dog, Beast.

"Are you sure you want to leave?"

"Yes."

I don't know what I expected him to say. Maybe I expected him to have some heartfelt response to my leaving after six years or my pouring out my heart to him about my passion to disrupt the media industry.

Instead, Mark remained as infuriatingly and lovably logical as he always is. "Why do you need to do this right now? You're about to have your baby. Take your maternity leave first and think about what you want to do."

Maybe it was the relief of being done with the Obama town hall or of getting my feelings off my chest with Mark. Or maybe it was the stress of what I had just lived through and pulled off. But whatever it was, the next morning—three weeks early—I went into labor with my son, Asher.

I was so sleep deprived for the next few weeks, I could barely even remember my own name, let alone think about career plans. But just a few weeks later, during one of my first solo outings out of the house post delivery (to Target—where else?), I received a phone call from Andrew Morse at ABC News.

"Randi, congratulations! You got an *Emmy* nomination for your political coverage."

There was no turning back.

After six years, which felt more like six lifetimes, I left Facebook.

To this day, people often ask me if I miss working at Facebook. Of course I do. But really, I miss a moment in time. My time at Facebook was like a wonderful vacation that can never be re-created or repeated and for which I'm forever nostalgic. There are a lot of things I miss about Facebook, but in particular, those early days were very special. I imagine most people working at start-ups feel that way, which is why it's so easy to catch the entrepreneurial bug in Silicon Valley. There's absolutely nothing like being part of a brand-new company, when you have no idea where the future will take you, when you can feel the energy pulsing off the walls, and you know that no matter where the road leads, you'll always feel connected to the people who are on that journey with you.

Immediately after leaving, I embarked on a yearlong speaking tour, during which I connected with people all over the world. It was refreshing and exciting after being in the Silicon Valley bubble for so long. I spoke extensively about my involvement at the intersection of tech and media. I outlined future trends in marketing. I dropped buzzwords like "social," "local," and "mobile" that audiences expected to hear, while also making sure to impart real, relatable advice.

One thing that really blew me away was that no matter where in the world I went, no matter what topic I was speaking about, everyone came up to me with the same, highly personal questions: How can I find out what my children are doing online? How can I ensure that I won't lose my job to someone younger and more tech savvy? How can I create a personal brand online to stand out more? How can I get my husband to stop using his iPad in bed?

And that's when I realized my calling. I had always been the "storyteller" inside Facebook—the one evangelizing how we can't forget that there are humans on the other side of the code, the visible spokesperson highlighting how Facebook was being used around the world to improve and enrich people's lives in exciting and unique ways.

Ironically, these same tools that delight people and create endless social and economic opportunity also keep us up all night and give us ulcers. In my year traveling and speaking, I realized that there are millions of people around the world who, although daily users of these technologies, feel overwhelmed, insecure, and confused about how these technologies are changing their lives, their families, and their careers.

I even saw in my own life that I was having difficulty balancing tech and non-tech moments. People talk a lot about work–life balance, but in today's society it's really more an issue of tech–life balance. As an entrepreneur, traveling the world speaking, while also trying to stay connected to my family, my friends, and my team, I reached a point when rather than owning a computer, a phone, and a tablet, those devices were owning me. I felt so much pressure to be always "on," always connected, that by the time I looked up, a year later, I had traveled to twenty-five different countries, made hundreds of new friends and business contacts, built a production studio, and launched a business. But somewhere in there I had forgotten to actually *live* my life without a device attached to my hand. I had forgotten how to just unplug and enjoy the company of those around me. I had forgotten how to be present in the moment.

So I set out on a mission: I wanted to help untangle all of our wired, wonderful lives.

Modern life is complicated. Keeping up with the latest apps, websites, tools, and gadgets is overwhelming. Parenting in the digital age may make you want to rip the hair out of your head. Navigating our professional lives, our love lives, our friendships, in an age where every action is public and documented, is confusing at best and career-ending at worst. But it doesn't have to be. Tech can fill our lives with meaning, rather than fear. Connecting to others can be empowering, rather than overwhelming. The gray zones can become areas of opportunity, rather than insecurity.

The Internet, social networks, and smartphones have given us amazing new tools and ways of communicating, collaborating, and living with one another. We can use those things to achieve change in our lives, relationships, careers, and communities. We can help redefine and revitalize art, culture, and entertainment. We can find balance and rediscover what it means to live in the moment. And we can use new technology to understand and solve some very old challenges that individuals and communities around the world have faced since long before Facebook, or anything like it, existed.

I'm going to share some of my personal stories, my struggles, my triumphs, and my excruciating and frequently humbling experiences encountering the new world of technology. I'm going to talk about what life will be like for digital natives—like my child—in the years ahead and about the challenges for this first generation of parents, who have children growing up entirely online, with every single moment of their lives documented and recorded. And I'm going to talk about how we can use technology for all the things that are important to us.

chapter 3

IT'S COMPLICATED

Sometime in late fall of 2012, I started to think a lot about technology and our modern lives: etiquette, relationships, identity, sharing. As I mentioned at the end of the previous chapter, I had just finished a yearlong speaking tour and was blown away at the number of personal questions people asked me about the role of tech and the changing communication in their lives—with their kids, their families, and their careers—no matter where I traveled or who was in the audience.

At about the same time, Bradley Lautenbach (my partner on the ABC News–Facebook debate back in 2008 and now my partner at our newly formed company, Zuckerberg Media) and I, along with Jeff Paik (our CFO), were weighing the pros and cons of a long slate of original content we wanted to create.

One day I was having a conversation with a woman about tech in the workplace when she suddenly started tearing up. She confided in me that she was going to lose her job to someone more tech savvy and asked for advice on how to get a better grasp on some of the latest tech.

That's when it hit me. If this person standing before me felt so tearfully insecure about tech and her life and her career, surely there were millions of other people out there who felt the same way. Tech didn't have to be overly confusing or complicated. When explained properly, in a relatable, approachable way, it can be amazing and life changing. I desperately wanted to help demystify tech and relieve people's fears, and in some ways, I felt like it was my responsibility to do so, the karma I needed to return to the world.

Bradley and I got incredibly excited about our new vision. Because Bradley had been producing for *Good Morning America* at ABC, he had a deep understanding of the morning television audience and how to produce content that was informative, entertaining, and relatable.

We knew that outside the tech world we lived in, people didn't really self-identify as "geeks" or "techies." If you ask most women what kind of content they want to read about in a magazine or blog, few will say they want more about technology. However, those same women will click on articles about technology at surprisingly high rates. We also saw that most people were more tech savvy than they gave themselves credit for and were very interested in content that helped them navigate their everyday lives.

Because tech is now such an ingrained part of our lives, it's really no longer "tech" content—it's simply *modern living*.

We saw that people were equal parts amazed and confused by technology. They loved downloading apps but were unsure how to monitor their children's usage or even how to just talk to their kids about the popular apps of the moment. Those who knew how to text message sometimes didn't know how to stop, even though they realized they were texting at times that might be deemed inappropriate. They loved sharing photos but often didn't know how to navigate the privacy settings around all that sharing. The gray areas of society were getting bigger and grayer every single day. And every new innovation added a new shade of complication.

It suddenly became very clear to me what we needed to do next. I knew that I wanted to help people understand how tech could be an amazing force for good in their lives, when used mindfully and properly. So, we decided to launch a website called Dot Complicated. The mission would be "untangling our wired, wonderful lives."

I cared deeply about the human side of tech. I had majored in psychology at Harvard and had been a longtime champion within Facebook of recognizing the *people* behind the computer code. I had spent the past six years working with people and companies to handle various issues related to tech and their lives. And I had acted as a spokesperson for Facebook in the media, when Facebook needed someone relatable and approachable to bridge the gap between tech and everyday life.

I guess, in some way, Dot Complicated had been a part of me all along. I suddenly felt renewed, invigorated. I felt like all the work I had done at Facebook—all the projects and innovations I had pioneered, all the mistakes I had made—had been leading up to this. It was time to address these issues around modern, digital living head-on.

With a newfound vision in place, Bradley and I decided to take off the week between Christmas and New Year's Day, so we could come back refreshed in the new year.

Part of being lucky means recognizing the signs that point out opportunity along the way. Even if those signs come hurtling at you in the form of an online uproar and public-shaming media spectacle at your own expense.

After all . . .

A successful woman is one who can build a firm foundation with the bricks others have thrown at her.

—modified from an original quote by David Brinkley

My family had gathered for dinner on Christmas 2012. My sister Donna is an amazing cook, and she was making Peking duck in honor of our family's modern Jewish tradition of eating Chinese food on Christmas. We're all so busy that it's a real treat to get everyone together. This was one of those rare years when we all just happened to be in town for the holidays.

After an absolutely delicious meal, my husband, Brent, took our son, Asher, home to get him ready for bed while I stayed a bit longer with the rest of the family to help clean up. Everyone was gathered in the kitchen, clearing dishes and drinking coffee, while Mark demonstrated the brand-new Poke app that Facebook had launched earlier that week. Through the app, you could send someone a message that would vanish in ten seconds.

This message will self-destruct in ten, nine, eight . . .

I would have had a lot of fun with an app like that in college, and I could see why this trend of "ephemeral messaging" was so popular among teens and young adults.

By that point, we had all downloaded Poke so we could try it out firsthand. Looking around, I thought it was funny that here we were, standing around the kitchen counter, and rather than speaking to one another, everyone was looking down at their phones, frantically texting and sending one another vanishing messages on the Poke app.

"Say cheese!" I said, pulling out my camera. "Pretend you're all sexting!" Everyone made funny pretend-horrified faces, and I snapped a quick photo.

I don't often post intimate family photos online, because I am a firm believer that you can (and should!) have meaningful relationships with people that you don't necessarily need to broadcast out to the world all the time. But this was an adorably tame, cute photo. So, I posted it to Facebook (under the friends-only privacy setting) and headed home to tuck my son into bed.

Of course, I knew there was a chance that picture would leak. I never post anything online that I wouldn't feel comfortable having reprinted on the front page of a newspaper. And this photo was the turducken of tech photos: Facebook family, using Facebook, on Facebook.

But I thought, of all nights, surely Christmas was a night when everyone was enjoying their own families and could appreciate our family photo without going, "OMG! Look! A photo of Mark Zuckerberg being funny with his family! I immediately need to blog this."

I had no idea what was in store for me.

About an hour later, my son was happily tucked into bed, and I was enjoying a mug of hot apple cider in my living room, just playing around online and procrastinating before going to bed. I took a quick glance at Twitter, and then did a double take. Someone had taken the family photo that I had posted on Facebook and posted it on Twitter. That meant that one of my Facebook friends had seen the photo pop up on Facebook, downloaded or taken a screenshot of the photo, saved it to his or her phone or computer, and then uploaded it to a totally different site. Because it was late at night, I was feeling a bit emotional from a nice evening with family. I fired off a response expressing my frustration.

Then I went to bed.

•　　　•　　　•

The next morning I woke up to what seemed like a national news scandal. I had dozens of text messages, several urgent missed calls from Bradley, and thousands of tweets. Every news station I flipped past was showing my family photo and talking about my Twitter exchange. Obviously, people were greatly enjoying the Schadenfreude of a Zuckerberg getting mixed up in anything that had to do with Facebook and privacy.

Gulp.

"Randi," Bradley barked at me, as soon as I returned his calls. "*Good Morning America* has texted me three times already this morning, asking for a comment from you on privacy and etiquette. Oh, and congrats on giving the media something to talk about for the next three days, during the slowest ever news cycle possible."

I just sat there amid a flood of incoming text messages from friends asking if I was okay, television producers calling me for interviews, anonymous people spewing venom at me on Twitter, Bradley asking me why he couldn't just go away for three quiet days of holiday without me stirring up a controversy . . .

But all I could think about was how "dot complicated" this whole situation was.

The entire media world was abuzz over the headline "Zuckerberg's Sister Caught Out by Facebook's Privacy Settings," but it really wasn't about that at all. I understood my privacy settings completely. This was about the gray areas of sharing, social conduct, and online etiquette.

Ironically, the downloads of Facebook's Poke app rose in the Apple store during the next few days. (See, I was still doing marketing for the company!)

More important, the whole incident made me even more passionate about starting a discussion about our modern, digital lives. Here I was, a living, breathing example of how tech could be a wonderful and amazing tool, but it could also get you into a lot of trouble. And I knew there were millions of people out there who could relate.

And with that, I dove headfirst into my new mission. I started doing regular segments for *The Today Show* on "modern tech dilemmas." I was quoted in the *New York Times* in a front-page article about the new voices of online etiquette. I launched my own e-mail newsletter, also titled Dot Complicated, with articles on how tech can improve your career, your love life, your family life, your relationships, and more.

And I started writing this book.

To understand why I think identity, humanity, and etiquette are so crucial to our modern lives and our relationships with tech moving forward, I think it's important to understand a bit about where we've come from.

The famous science-fiction writer Arthur C. Clarke once said, "Any sufficiently advanced technology is indistinguishable from magic." He was right. New technology is a kind of magic, and today we can do things with ease that were impossible just a few years ago. Like magic, each new innovation has advanced our society and our potential. Of course, the seductive glow of these magical devices can also blind us to some of their downsides and side effects.

My journey began in the magical land of Dobbs Ferry, New York.

They were the glorious hypercolored-T-shirt-and-Umbro-shorts-wearing days of the early '90s. I must have been nine or ten years old at the time. One day, after I returned home from school, my father called me into his dental office, which was located on the bottom floor of our home.

"I have a surprise for you," he said.

I better not need a filling, I thought.

I went down to the office, and there I discovered the surprise: a large beige box, on a little trolley with wheels, plugged into the wall. There were a couple of blinking lights on the side.

I was unimpressed. "So, what does it do?"

My father duly explained. "We can take a photo of you and put it into the machine, and then we can take another photo of someone else and swap your smiles. So, now when people come to see me they can choose the smile they want and know what it'll look like."

I was so excited that I insisted my father show me the machine right then and there. We ended up scanning and swapping my smile with my mom's.

Of course, I couldn't immediately share the picture with thousands of people, like we can today. My dad printed out a copy and I brought it to school, showing a few friends, but that was it. Then I promptly placed the image inside a shoebox of photos beneath my bed and forgot about it. If I could have shared it with thousands of people, would I have done so? Should I have done so? Luckily, things were less complicated back then. And besides, I looked pretty weird with someone else's smile.

I was fascinated with that machine. I used to try to find ways to sneak down to the office when my friends were over and we would happily play a few rounds of smile swapping. Sometimes I would get caught and then I would get an earful. But it was totally worth it for a chance to play with magic.

In seventh grade, I got my first telephone. Because mobile phones weren't yet prevalent, it still had a cord attached to it. And okay, it wasn't even entirely mine. During the day, it was my father's office line. But after the dental office closed for the day, I was free to claim the line. Now I could spend evenings and weekends talking with friends for hours! Obviously, we talked about all the important topics: boys, movies, Nirvana, and Ace of Base. Sometimes I'd also have unexpected conversations with people looking for some dental work. Those conversations were less interesting.

At about the same time, I had my first encounter with the Internet. Computers weren't new in our house. For as long as I could remember, my dad had a couple of old machines in his office, an Atari from the 1970s and an IBM PC he'd bought at about the time I was born. I never really touched them; my father used them for storing patient records, doing office correspondence, and other serious stuff. But in the mid-1990s, my parents got a computer for my siblings and me to share.

That first computer was big, slow, and often difficult to use. Listening to the computer cringe and wail as it dialed up to the Internet was a tedious experience, one made even more tedious by getting disconnected every time someone in the house picked up the phone. But logging on to AOL for the very first time and being able to send e-mail, search Grolier's online encyclopedia for school projects, or instant message with my friends was life changing.

In my senior year of high school, I got my first cell phone. It was a big, unwieldy Nokia 5110 in a lime-green case. It looked like and felt like a brick, and that's what I called it. But, again, it opened up a new world of amazing possibilities. For the first time, the ability to reach someone and be reached wasn't tied to a physical location. I no longer had to wait outside the east entrance to the mall for twenty minutes because my friend was at the *west* entrance.

If you're lucky, you skipped that whole dial-up Internet phase. Maybe your first phone was a smartphone. No playing Snake for hours at a time on a tiny monochrome screen. But however those moments played out, they all tell a common story. Technology and connectivity are becoming more advanced every day.

Once it was cool to have your own landline. Now five billion people around the world have their own mobile phones, and about a billion of those are smartphones with access to the Internet, e-mail, and all the amazing apps that are available today. We've gone from a world where a decade ago Internet access was a novelty to one where 2.4 billion people are online. And millions more join every day.

Devices are faster, cheaper, and more powerful than anything we could have imagined even a few years ago. In Silicon Valley, every geek is familiar with Moore's law, named after the founder of Intel, Gordon Moore. Back in 1965, he predicted that microchips would double in power roughly every eighteen months. That prediction has remained true. Also, as computers have continued to get faster every year, the cost of devices has steadily fallen.

This trend is what has allowed so many more people to get their hands on computers, smartphones, and tablets over the last decade, and why these devices keep appearing in ever more places. Moore's law is what has led to billions more people getting phones and hooking up to the Internet, and why a phone today is now a hundred thousand times more powerful than the computer on the Apollo spacecraft that took men to the moon.

Plus, people share more online every year. A few years ago, someone came up with the name for a technology trend designed after our own family! "Zuck's law," so-named for my dear brother, says that the amount of information we share in the world doubles every two years. In the time it takes you to read a single page of this book, another hundred hours of cat videos, hilarious skateboarding dogs, and many other, er, valuable pieces of content will have been uploaded to the web. Over a million tweets will have been shared.

Some of them will actually be read. And over sixteen million pieces of content will have been posted on Facebook. That's a lot of photos of people's lunches.

Of course, there are limits to this trend. At some point, people simply can't handle any more information—there are only so many hilarious cat videos and cute baby pictures we can look at. But you get the idea. The scale of what's being shared is almost beyond our comprehension. Here's an incredible thought: according to a *Digital Universe* report published in December 2012 by the analyst firm IDC, there are more pieces of digital content in the world today than there are grains of sand on every beach on Earth. Wow.

And none of these trends is slowing down.

The whole world is getting connected. You can get online anywhere. You can connect to the Internet on top of Mount Everest. You can find a signal at the International Space Station. Many astronauts have built up online fan followings by posting mind-blowing photos of the Earth from orbit. Talk about a Kodak moment.

So, in less than four decades we've gone from talking about connecting the world to *actually* connecting the world. And our expectations of connectivity are becoming a lot more demanding.

While growing up I remember how exciting it was to get online for even a few minutes and how lucky I felt to enjoy a good thirty minutes of chatting with my friends on AIM (AOL Instant Messenger) without one of my family picking up the telephone. Then I would wail frantically for them to hang up.

Today, we expect to be online all the time, and we expect to be reachable everywhere. We usually are. A Morgan Stanley *Internet Trends* report shows that over 90 percent of people keep their mobile phones within three feet of them, twenty-four hours a day. A May 2012 Harris poll in the United States found that 53 percent of people regularly check their phones in the middle of the night after they've already gone to bed, and a surprising and slightly disturbing number of people check their phones while on the toilet. (I don't imagine most wash their handsets afterward. Think about that the next time someone hands you his or her phone and asks you to take a photo with it. You're welcome.)

This online-all-the-time mentality pervades every area of our lives. Two 2012 surveys, one from Yahoo! and the other from Gazelle, revealed the following eye-opening data:

25 percent of women would give up sex for a year to keep their tablets.

15 percent of all survey respondents would give up their cars to keep their tablets.

Nearly 15 percent of all survey respondents said they'd rather give up sex entirely than go for even a weekend without their iPhones.

A 2012 TeleNav survey asked people which of life's "little pleasures" they would rather do without for a week, instead of parting with their phones:

70 percent would give up alcohol.

21 percent would give up their shoes.

28 percent of Apple product users would go without seeing their significant others; 23 percent of Android users agreed.

A recent study from McCann Truth Central claimed that 49 percent of married moms would give up their engagement rings before they would part with their mobile phones. And a 2012 study from

Harris Interactive revealed that 40 percent of people would rather *go to jail* for the evening than give up their social media accounts.

So, this is the world we live in now. Technology is almost everywhere and has come to dominate our lives. So much so, in fact, that we're starting to see people yearning to be less connected and trying to implement rules, structure, and discipline in both their own and their families' lives, to ensure that all this connectivity does not come at the expense of relationships, skill development, and manners.

It's going to become increasingly important to find that balance, because in the next decade we're going to see something even more extraordinary. Everyone and everything will be connected. There will be no division anymore between online and offline.

Beyond people, we'll see objects, our environments, our homes, our clothes, and our cars come alive with data. One of the most popular Silicon Valley predictions is of a future with an "Internet of Things"—a world where our cars, kitchen appliances, and even shoes are connected. We're well on the way to seeing that become a reality. According to a Cisco study in April 2011, there are between ten and fifteen billion connected devices in the world today, but by 2020 that number will have reached fifty billion.

A few months ago, I had a cute, hilarious, and sort of terrifying moment with my son, Asher, which showed me what tomorrow might look like. Asher has come to know and love *Barney and Friends,* the children's TV show about that famous purple dinosaur. Asher loves to sing and dance along with the characters, and he could easily watch the show for hours on end, if only I would let him. One afternoon I was working on my laptop while Asher was playing with his toys on the rug. He got bored after a while, and out of the corner of my eye, I saw him staring up at a photo frame on the bookcase. It was a photo of my parents.

"What is it, love?" I asked.

Asher looked at me and then pointed at the frame. "Booney?" he asked.

For a moment, I didn't realize what he was talking about. But then I got it. Asher had come to realize that content always flows from screens. On the big TV, he can get *Barney*. On my iPad, he can get Barney. So, surely a photo frame must also be able to conjure up his favorite television character.

I laughed. Asher looked disappointed.

But then I thought, *He's absolutely right. Why shouldn't he be able to watch Barney on that photo frame? For that matter, why shouldn't any device be able to show us any information we want?* One day he'll be right, and already his child's logic shows just how intuitive and obvious this future will be. Every piece of glass will be a screen, and every screen a portal to another world of information, content, ideas, and entertainment. There's absolutely no reason there can't be a purple dinosaur in every frame.

As a new mom, that's both exciting and utterly terrifying. It's hard enough to balance screen time versus non-screen time as it is. What happens in a world where everything is screen time?

The future will be a place of infinite possibilities. No one will ever need to sneak into his or her dad's office again for a chance to experience the magic. But as we get more and more connected, it's also going to become increasingly important to know when to step away, when to focus on the people and places around us. A world where every object is a screen means a world of endless access to information, but it also means a world where we risk jeopardizing our relationships with loved ones if we don't look up from that screen from time to time.

Our definition of "magical moments" may change to become

those increasingly rare simple moments when nobody is connected, and there is no magic whatsoever.

What's the upside? We're more connected.

And the downside? We're more connected.

Technology has altered every aspect of our lives, from our relationships to our families to our careers to our love lives. It's changed how we celebrate birthdays, how we announce major life news, how we define friendships, and how we demand customer service.

With smartphones, and the cameras built into them, friends and family can share all the most important moments in their lives with one another as they happen. In June 2011, a Pew Research Center survey of over two thousand American adults found that Facebook users have stronger ties with their closest friends, find it easier to get support and advice from people, and are more likely to stay in touch with "dormant ties," old friends from high school or college, or people who live far away.

Grandparents can see the face of a newborn grandchild from thousands of miles away through the lens of a webcam and via video calling. In fact, research published in 2012 by Dr. Shelia Cotten at the University of Alabama, Birmingham, showed that seniors who used the Internet were about 30 percent less likely to be depressed than seniors who didn't.

Colleagues can have virtual face-to-face meetings with people working in offices anywhere in the world—there's no such thing as a remote office anymore.

Friends can capture every moment of a dinner party through photos and make those photos beautiful with professional-looking edits, filters, and borders.

That same ease of communication might also mean that you get an informal Facebook message on your birthday, instead of a phone call; that you might get an e-mail from the person sitting right next to you at work, instead of an actual conversation; or that everybody at your dinner party might be so busy taking photos and making them look nice, that they're no longer paying attention to anybody else. We can miss important moments if our heads are constantly buried in those phones.

Today, everyone is a broadcaster as well as a receiver. In the past, we were all just passive consumers of information. Creating content was reserved only for the rich and powerful, who controlled and ran large media companies. But now, each of us can generate and share as much as we receive.

When I wrote for my high school newspaper, we had a staff of twenty students working diligently to produce a paper with a total circulation of about a thousand readers. Nowadays many of us can reach over a thousand people with a single tweet, photo, or Facebook status update.

Now everyone is a media company. We can use technology to make our voices louder and heard in more places. When people come together online to raise their voices as a chorus, truly spectacular things can happen. In 2008, an unemployed twenty-one-year-old engineer in Colombia named Oscar Morales set up a Facebook page protesting the FARC, a terrorist group in his country. FARC had been kidnapping people, planting bombs, and terrorizing innocent people for years. One evening, sitting on his computer, Oscar read the news of another attack. Out of frustration, he created the page and named it optimistically "One Million Voices Against the FARC!"

He didn't expect to get nearly that many people. But then something incredible happened. Within four hours the page had fifteen hundred members. The next day it had four thousand. By the end of the week it had a hundred thousand.

Stunned by the unexpected success of his online movement, Oscar did something he had never expected to do. He called for a nationwide day of protest against the FARC.

One month later, twelve million people marched in two hundred cities in what became the largest demonstration against terrorism in history. And in the end, the wave of political pressure the marches set off greatly contributed to sending the Colombian government and the rebels back on the road to peace talks.

What Oscar did was amazing and courageous. But not unique. Today, all over the world, in coffee shops and backrooms, in dorm rooms and town squares, other brave but otherwise completely ordinary people are using the tools of technology to stand up and advocate for change in their communities. In the hands of young, idealistic, well-organized activists, services like Google, Facebook, and Twitter become more than just a set of apps or Internet destinations. They become the tools of change. They become pathways to freedom.

That's exactly what we saw happen with the use of the Internet in places like Egypt and Tunisia during the Arab Spring, when large numbers of young, motivated activists took to the streets there and in countries throughout the Middle East and North Africa to demand greater social and political freedoms, and it's what we continue to see in Russia, China, Iran, and many other countries.

And change goes beyond just the political. The Internet is playing a meaningful role in creating jobs and improving economies. Today, even as the global economy remains fragile, the Internet is emerging as an indispensable force for growth, jobs, and opportunity, through both the rise of giant new web companies and the realization by the older existing companies that the Internet is essential to their businesses. In a 2011 McKinsey Global Institute report, *Internet Matters,* researchers found that for every job lost to the web, two more were created by the web. These figures are only going to increase as more people and services go online.

The economic benefits of connectivity aren't confined to advanced economies. The World Bank's *Maximizing Mobile* report in 2012 showed how potato farmers in India managed to increase their incomes by up to 19 percent by using mobile apps to support their businesses, and banana growers in Uganda by 36 percent.

When disaster or tragedy strikes, the Internet can help save lives, mobilize aid, reunite loved ones, and assist the most vulnerable members of society. After the Japanese earthquake in 2011, hundreds of thousands of people took to Facebook, Twitter, and Google to find missing friends and family, and billions of dollars in online donations poured in from around the world to relief organizations working on the ground.

The Internet allows, and encourages, information to travel faster and farther than ever before. That means positive information travels quickly. If I have a particularly great experience at a restaurant or vacation spot, I can recommend it to everyone in my network. It also means negative information travels just as quickly. If I have a terrible experience somewhere, all my friends will immediately know about that as well. Therefore, businesses need to double down on their customer-support efforts. A world where everyone is a media company is a world where businesses can no longer afford poor customer experiences.

I once overheard a major Hollywood film executive say, "Social media has ruined our ability to release bad movies. And we need to be able to release bad movies to stay in business." It used to be the case that a really bad movie could still have a great opening weekend, because it would take word of mouth a few days to spread. But in the age of Facebook and Twitter, a movie can be dead in the box office just hours after it opens.

But just because we have a megaphone doesn't mean we need to shout from it all the time. If we're constantly crying "Wolf!" nobody will take us seriously. As a society, we need to accept the gift we've been given and realize that it comes with a set of responsibilities. When used thoughtfully and mindfully, we can expand access to knowledge and information, demolish old barriers to understanding, and give a global voice to those who were once voiceless.

In the next decade, another three billion people will go online, mostly on mobile phones. In our homes, physical objects, such as recipes, photo prints, and receipts stuffed in drawers, will all be replaced by apps. Our doors, alarms, lights, thermostats, closets— all will be controlled with the touch of a button. On the road, cars will become platforms in and of themselves, acting as our travel guides, our virtual assistants, and eventually, with self-driving cars, even our drivers. We'll be able to monitor our health daily, through wearable health devices and smart clothing. How we consume content, how we educate our children, how we view our possessions— everything is changing at a breakneck speed.

Once upon a time, all this was science fiction. Now it's science fact.

 SELF

Me, Myself, and Identity

It's exhausting to put on an act—to be somebody in one situation and somebody else in another. Others may advise you to change who you are, to pretend to be someone else to get ahead, to "play the game." But unless you're committed to putting on an act all the time, it can be difficult to keep track of who you are in each situation and who knows you in what capacity.

Being authentic means you might not win a popularity contest anytime soon or be best friends with everybody. But, personally, I've always found it easier to sleep at night when I was true to myself that day.

Luckily, the Internet is beginning to catch up. Over the past decade, we've seen a shift from people using anonymous screen names online to more often using their real names, their real identities. This change in behavior, and the increased comfort with putting more information about ourselves online, has helped drive a great deal of innovation and change.

Remember your first, probably embarrassing, e-mail address or AOL Instant Messenger screen name? I do. Mine was Peggy42st, which I picked because I had just played the part of Peggy in my high school's production of the musical *42nd Street.*

I remember having a whole discussion with my mom, right after I was accepted into Harvard University, during which she assured me that people would be normal and not stuck up or full of themselves. Immediately after that conversation, a future classmate tried to strike up a conversation with me on AIM. His screen name? igot1600.

So much for your theory, Mom.

Remember the effort you'd make to choose the perfect IM profile picture, or the time you invested in crafting the perfect "away" message? I admit that I spent way too much time selecting vague but meaningful lyrics from the latest song I was obsessed with. Plenty of times I would announce my presence online with "I believe I can fly," "I saw the sign," or "I get knocked down."

But, look. I'll come clean. And this may be hard to believe, but I wasn't *really* Peggy from *42nd Street.* I was Randi from Dobbs Ferry. I could *call* myself Peggy online, but I wasn't Peggy. In the early social web, that didn't matter. My friends knew Peggy was Randi and didn't get too confused about the whole thing.

But it would be weird if I had a profile of Peggy as my Facebook profile or used igot1600 as my screen name today. That's not just because the SATs are no longer out of 1600 or because I've grown up. It's because the Internet has grown up too.

Back when the web was mostly about accessing information, we experienced the Internet by typing keywords into search boxes, looking for just the right bright blue link to get us the latest news, celebrity gossip, movie listings, maps, and school research. Search engines offered us answers at the push of a button.

In the past decade, the Internet has grown beyond just finding information to connecting with *people*. Now we can benefit from the wisdom of our friends.

I remember the exact moment, on a Virgin America flight between San Francisco and New York City, that my friend gave me a stern talk about how I should change my Twitter handle from @randijayne to @randizuckerberg so that people would be able to find me.

When we use our real names and identities online, we can easily find and connect with friends, family, and colleagues no matter where we travel in life. We can show ourselves to prospective employers as the people we really are, with our résumés and career histories on display to the world, allowing us to find new opportunities and livelihoods. We can interact with businesses as more than just random website visitors and benefit from products, offers, and services that are more personalized, relevant, and useful. Smart advertising systems that better understand our interests and personal information can suggest things we might actually want to buy or even allow us to just have fun, like serving up ads for dude ranches when we title our e-mails "Dude" in Gmail.

I'm often asked to talk about Facebook's early success, how it was able to gain so much momentum so quickly. Of course, most of this was due to the excellence of the site and the product itself, but I think a big part of Facebook's early success was due to people using their real names on the site.

Right from the beginning, there was a culture on Facebook of this. While it was definitely not the norm on other social networking sites at the time, such as MySpace or Friendster, people on Facebook felt comfortable using their real first and last names because they had to be authenticated using a .edu e-mail address, which meant that the only people who could see them on the site were the same people they would run into in class, in the dorms, or at parties on campus anyway. This established a level of trust early on and meant that the connections formed on Facebook were more valuable and more authentic than on other sites. More important, it meant that people tended to behave thoughtfully and truthfully. People were much less likely to be nasty or make up blatant lies about themselves, when they could be easily called out on it.

On a grander scale, when we speak as ourselves online, our voices carry further, with greater authority. All those amazing examples of online movements that have created massive social change were only made possible because courageous individuals inspired others to action by taking a stand as themselves. Identity can be more than just consuming inspiring quotes or cute photos from a friend. It can mean the difference between simply believing in an idea or a cause and standing in the way of tanks.

In a world where we are the same people online and offline, we'll know more about the people, celebrities, politicians, and, really, everyone we interact with.

I'm a passionate believer in the power of authentic identity. So, I've been quoted as saying that anonymity on the Internet had to go away, as a way of unlocking all the great benefits of an authentic web. I also think it will help in the fight against online bullying. It seems obvious to me that people tend to behave better, and abuse others less, when their behavior online is out in the open for all to see, attached to their real names and a picture of their faces.

That's not to say there's no more room for witty-and-weird or anything-goes online screen names, or that no meaningful relationships can arise from online interaction via avatars. In some communities and countries, or when discussing certain extremely sensitive topics, the cloak of anonymity can be necessary simply for personal safety and privacy. Activists and campaigners who are working for change in societies around the world may need to organize and meet in secret online. People asking questions about their health, or asking questions as victims of a crime, may not want to do so under their real names. But these are the exceptions to the rule. Just as there are perfectly good reasons why we expect people to use the same, real names in their daily lives, the Internet will be a better place when everyone consistently uses their real names.

And as the Internet becomes defined by our authentic selves, who you are online will reflect more of who you are offline. Today, it's becoming hard to say you actually went anywhere or did anything if you don't have the photos or the location check-ins to prove it. "Pics or it didn't happen," as the saying goes.

We are both the artists and the curators of our online one-person shows. Our digital selves are quickly becoming reflections of our actual selves. In a 2013 Cambridge University study that analyzed the Facebook profile data of about fifty-nine thousand Americans, researchers discovered that, based on likes alone, they could accurately predict a person's gender and ethnicity 95 percent of the time, whether a person was a Republican or Democrat 85 percent of the time, whether a person was Christian or Muslim 82 percent of the time, and even whether that person's parents divorced before he or she was age twenty-one about 60 percent of the time.

Authentic identity is what enables Florence Detlor, a 101-year-old grandmother in Menlo Park and one of the world's oldest Facebook users, to stay up to date with the lives of her grandchildren, friends, and family.

It's what enabled Aaron Durand, a Twitter user from Portland, Oregon, to save his mother's independent bookstore from going out of business. When he posted a single tweet asking for help, promising to buy a burrito for anyone who bought more than fifty dollars worth of books, the message ended up being shared by hundreds of people. His offer drove so many new customers to the store that it became profitable and managed to stay open.

And authentic identity is what reunited João Crisóstomo, a sixty-eight-year-old New Yorker, with his long-lost friend Vilma Kracun. João had known her while working as a waiter in London in the 1970s, but after he moved to Brazil and then New York, he lost touch with her. Four decades later, on Valentine's Day in 2011, he received a phone call from a mutual acquaintance: Vilma had been found on Facebook. They soon made plans to meet in Paris, fell in love, and were married in April 2012.

Of course, all change comes with growing pains. Online identity can be a challenge to manage, especially when you are using many different social networking sites, banking sites, dating sites, job-search sites, and more. Entire industries have popped up to help people store and manage their online accounts and improve their online reputation, and to help websites show up higher in Google search results.

And then there's the issue of privacy. While the positives of being your real self online far outweigh the negatives, it's always scary to open up more of yourself online. But before we scream bloody murder, we have to collectively take a deep breath and try to figure out if we're uncomfortable with something because it's truly bad or because it's new and we are resistant to change.

One example I like to reference is the introduction of caller ID on our phones. When caller ID was first introduced, people were up in arms. How dare our privacy be invaded! How dare other people be able to see that we are the ones calling them! But now think about a world without caller ID. I don't know about you, but when I get a call from a number I don't already have programmed into my phone, I usually let it go straight to voice mail. Caller ID has been a tremendous net benefit. Often change, even though it feels uncomfortable at first, can be for the better.

So, let's embrace being our real selves all the time. It's exhausting to try to be someone different in different situations. When we're authentic online, we can connect with others as we really are, and vice versa.

Welcome to the Gray Zone

Of course, for all its amazing benefits, the Internet can get you into lots of trouble. That's the thing about this "sufficiently advanced" technology. It may be magic, but we're not magicians.

In the early days of Facebook, mostly everything we did on the site was modeled after some parallel thing that existed in colleges around the country. There were features within the profile for listing your classes and your spring break plans; even the "wall" (now part of the Timeline) back then was more of a whiteboard, with the capability of displaying only one message at a time. By the same token, instead of ads on the site, Facebook had "flyers."

One night, working late on a PowerPoint deck for the marketing department, I needed a picture of a flyer live on the site. A colleague and I quickly drafted a convincing flyer for a fake rush party at a Stanford sorority the following day (we used the name of an actual sorority she had heard about through a friend), posted it to the network, and took a screenshot. With my presentation complete, I shut down my computer and went home to crash. I was exhausted from the long night of work.

The next morning, I awoke with a horrifying realization. I had forgotten to take the flyer down.

Sh*$%!

Of course, we immediately took down the post, but it was too late. Sorority pledges had begun to gather, searching for a pledge party that didn't exist. More and more of them came, by the dozens. Moreover, it was midterms week on campus, during which time sororities and fraternities are banned from holding formal events. The accidental flash mob dispersed only when a bullhorn-shouting security guard insisted, "No party. No party!"

The sisters of Kappa Kappa Something were pretty unhappy with Facebook after that. It was my task to clear things up with the administrative office before the sorority was banned from campus. As a valiant gesture, one of our engineers kindly offered to host a sleepover at Facebook for all the girls, "to make it up to them." Somehow, I didn't think that was such a good idea. But we managed to square things with everyone in the end.

I learned a couple of valuable lessons that day. For one, Facebook ads clearly worked. More important, it was a huge aha moment for me. Until then, I hadn't witnessed firsthand the incredible impact a simple online action could have in mobilizing people offline. One click of a button might seem minor, but in reality, it could be extremely powerful. In some ways, I credit that experience for guiding my thinking in many of the projects I went on to lead, involving politics, nonprofits, and pop culture. In this new era of social media, when everything is broadcast to the world, even simple messages can have potentially far-reaching consequences.

There are a lot of pretty bad things that the "online you" can do to the real you. Your emotional texts, sent without thinking in the heat of the moment, could become widespread. You could accidentally post an intimate photo taken for your eyes only. You could forward someone an e-mail without seeing that there's something that person shouldn't see at the bottom. You could mistakenly reveal someone's big news before that person is ready to share it.

There was a time when, if you wore the wrong kind of clothes to middle school, at worst the only people who would notice would be the clique of mean girls at the cool-kids table. Now, the cool-kids table is everywhere.

The Nuanced but Important Difference Between "Private" and "Personal"

Before the Internet came along, we could categorize our information in three simple ways: public, private, or personal.

Public information is exactly what it sounds like. This is everything you're totally cool with people knowing or having access to, or winding up on the front page of a newspaper.

Private information includes the things that you would only tell your lawyer, therapist, doctor, spouse, diary, or absolutely no one at all.

But then there's personal information—the category in between, full of complicated nuance. This includes things you might tell your friends but you probably wouldn't share with strangers.

I post family photos on Facebook all the time for my friends to see. I see photos of my friends' kids, weddings, and families. None of these photos are private per se. It wouldn't ruin anyone's life if these cute, harmless photos wound up in the press. But these images are certainly personal, which means my friends trust me to behave appropriately when I see them.

It's generally pretty easy to identify information as either public or private. But when it comes to personal information—that middle ground between something that's okay to share slightly outside your immediate circle but not with absolutely everyone—you enter a bit of a gray zone.

Before the Internet arrived, the gray zone was there, but it was much smaller. You could be fairly certain that if you showed your vacation bikini pics to your friends, it didn't mean your aunt and your aunt's friends and some random guy you once went to high school with were also going to see, distribute, and comment on them. Your friends would understand the nuance and context of what they were seeing and know not to share personal material with the outside world. Sitting around the coffee table looking at your photo albums, you controlled the distribution of your information.

But online you don't have that luxury. Online you have private and you have public, and the entire concept of personal information has vanished.

This causes problems.

I learned this lesson when some photos of my bachelorette party didn't stay in Vegas like they were supposed to. The photos weren't scandalous; it was just me and a few friends hanging out by the pool. I didn't think twice about posting them on Facebook, but from there the photos found their way onto a Silicon Valley gossip blog, where I became the subject of some uncomfortable discussions.

It was embarrassing but not awful. Nobody was nude or inappropriate—just a bunch of girls having fun at a routine bachelorette party. Of course, I would have preferred that the photos not wind up on the broader web, but nobody cried about it. It didn't ruin anyone's life or career. To this day, I still don't know who saw the photo album and spread it more widely. Hundreds of my friends managed to see the photos and categorize them appropriately as falling into the personal zone. But the moral of the story is that it takes only one friend who doesn't understand or respect the difference for the personal zone to go away.

When personal information goes public, it's really hard to know what to do about it. There are two schools of thought here. Some people feel that it's okay to ask for the information to be taken down. But many more will argue that you should stay quiet and hope it doesn't spread, that when you object to the publication of information you previously considered to be personal, it only adds fuel to the fire and your protest will, ironically, make it even more likely to be seen by more people. This is known online as the "Streisand effect," named after a certain popular singer's famously failed attempt to have a photograph of her private home removed from the Internet.

This whole situation needs to be fixed. It can't be that we're going to have to adjust to a world where we cannot share anything but our utmost public and sterile information. Sharing the personal stuff with others is an essential aspect of what it means to be human. If our online lives are to be as fulfilling as our offline ones, and if those two lives are to be fully integrated, then as we go forward we need to find a way to bring back personal information online. We must be able to post some pool pics without the whole world finding out, even if one of our friends is feeling a little overenthusiastic with the share button that day.

Did you ever wonder why, in movies about the Old West, despite how hot it must have been, the frontier folk wore a lot of heavy suits and always spoke in a kind of flowery, polite way? Why, in the middle of nowhere, would people care so much about how they spoke or whether their clothes were dashing?

I think this featured heavily in movie producers' depictions of the era not just because it was fashionable but because that kind of behavior played an important role in maintaining order in a strange new environment. The Internet is the latest technological improvement bringing us into unfamiliar territory, and as with previous frontiers, we're going to need to grasp something that our parents, our teachers, and our communities probably spent a great deal of time instilling in us as we grew up, something known simply, for lack of a better word, as etiquette.

Repost unto others as you would have them repost unto you.

It's still the Wild West online, and we need new guidelines for the digital era. Even though so much about the way we communicate has changed, certain basic rules of decency and civility haven't changed and, in fact, may be needed now more than ever.

Tips for Achieving Tech—Life Balance in Your Own Online Identity

Be Your Authentic Self—Online and Off
Using your real name and identity online definitely comes with its fair share of challenges, but the pros—from the ease of connecting to others to building your personal brand—far outweigh the cons.

One Small Click of a Button, One Giant Heap of Repercussion
Even the seemingly smallest post online can have far-reaching consequences and be eternally searchable. So please, please think before you post!

Don't Mess with Sorority Girls
This is more of just a general life lesson, actually.

Drawing the Line Between "Public" and "Personal"
There is an important distinction between these two, and at some point the concept of "personal information" stopped existing online. Let's each do our part to be a conscious digital citizen

so we can all feel more comfortable posting online. Also, just because one friend doesn't respect the privacy of your personal info doesn't mean you should stop posting. But it might mean hitting the "unfriend" button on a serial offender.

Repost Unto Others as You Would Have Them Repost Unto You

The golden rule of life applies to social media. Treat other people's updates as you would want your own to be treated, and we can all have a "woo"-filled world.

As is the case with any technology or tool, it's up to *people* to make the most out of the tech, to utilize it in a way that enhances their lives and relationships instead of detracts from them. Technology can make our lives more interesting, but it's not going to solve all our problems, and as we've seen, it will probably even create a few.

It's a common complaint that people are beginning to see the world only through the lenses of their camera phones, as if those screens were more "real." At concerts, I've seen well-meaning souls watch an entire show play out on the shaky screens they're holding above their heads, rather than the stage in front of them. I've seen people so busy Instagramming a moment that they miss truly experiencing it.

Keep in mind that an Internet with real identities and standard practices of behavior doesn't have to be a boring place or a police state. It just needs to mirror how you'd behave in a similar situation offline. If you're at your parents' house for Thanksgiving, it's generally inappropriate to down a row of tequila shots, slap your parents' friend on the back, and yell, "Woooo! Spring break!" Of course, if you're celebrating Thanksgiving in Cancún and crazy Uncle Al is in an especially "festive" spirit, then by all means woooo away. A call for mindfulness, compassion, and etiquette online, which mirrors the standards expected of us in civil society, is *not* a call for a woooo-free world. The point is that as our online lives have become inseparable from our offline lives, we need a set of rules, taboos, and guidelines that recognizes there are real people using their real identities on the other side of the screen.

Some things are cool on spring break that are not cool on Thanksgiving Day. There may be some awful places on the Internet, which don't deserve a mention here. You may see some things you don't want to see, things for which your retinas will never forgive you and that may be acceptable only in those contexts, within of course the bounds of legality. But that doesn't mean there are no standards of behavior to be had *anywhere* online.

We have to get smarter not only about what we publish but also about what we, as the recipients of our friends' information, do with potentially sensitive material posted by others.

Above all, you need to be careful who you choose as your friends, whether offline or online. As my Christmas Poke photo story shows, there's no privacy or security setting in the world that can save you from a friend's bad judgment.

We can be our real selves online and off. We don't have to be afraid to share. It doesn't have to be so complicated. And we can leverage technology in many positive ways to make a real and meaningful impact on the world.

We are truly the most empowered generation in history. Technology allows us to communicate, collaborate, and understand the world around us in ways unthinkable even a few years ago. With this new power, we can solve age-old problems and create new opportunities for everyone.

All you have to do is be yourself.

 FRIENDS

Closer to Friends, Further from Friendship

T he call was from a friend of a friend who had been introduced to me over e-mail by another friend of a friend who had not asked my permission before making the introduction. But I digress. This friend of a friend owned a business and was keen to pitch me the services his company could provide to our new production studio in Menlo Park.

I tried to be pleasant in my initial e-mail response, while also avoiding setting an exact time for a call, hoping I could weasel my way out of it. But a couple of days later, the e-mails started.

"HEY RANDI!!!" the first message began. "I'd LOVE to find some time to catch up tomorrow on the phone. I think we could do SO MUCH together."

At the time, I was juggling a bunch of crazy deadlines on two huge client projects, traveling between New York and California, and trying to maintain my sanity. I had barely seen my own family let alone had the time for someone completely random. I wrote back and vaguely promised to get in touch soon. *Preferably after he fixes the caps lock key on his keyboard,* I thought.

"Sure thing!" he wrote back instantly. "How about the day after tomorrow?"

Oh boy.

Over the next couple of weeks, we traded a couple dozen e-mails. He was easily one of the most determined and oblivious people I had ever met. One thing you should know about me: I very rarely use my phone as a phone. I e-mail and text up a storm. I have hundreds of apps. But I'll go weeks without checking my voice mail. (My voice-mail message even says not to bother leaving me a voice mail.) So, with that in mind, I really didn't want to fit another call into my already jam-packed schedule, and I didn't understand why he couldn't just e-mail me his request. But he kept coming up with his relentless requests for a phone conversation, and eventually I caved and scheduled an early evening call.

At the appointed hour, the phone rang and I began to hear the pitch. I let him talk for about twenty minutes. The pitch was long-winded. Twenty minutes became thirty. The minutes kept ticking by. I was tired and needed to get home to relieve Asher's sitter. Eventually I decided it was time to end it.

"Well," I announced, "it was so good to talk—"

"Sure thing!" he interrupted. "But I wonder if we might continue the conversation in person? Can we meet tomorrow?"

Oh geez. Two dozen e-mails to get to a phone call, which had really just been a prelude to a meeting?

"I think we covered everything in detail, don't you?" I replied.

"Ah, well, um . . ." he mumbled.

I made my farewells and hung up as forcefully as I could. Tapping the little red hang-up button is nowhere near as satisfying as hanging up used to be. I closed up the office. Everyone else had left for the evening. I made my way to the parking lot.

As I climbed into my car, I got a text from Brent. "Got home first! Asher was exhausted so I tucked him in already. No need to rush home."

I remember sitting there for a moment in my car, in the empty parking lot, suddenly feeling sad and angry. A random person had taken nearly an hour of my day, and I had missed tucking my son in to bed that night.

Even as technology provides new ways to connect with the people who matter most to us, often it is the ones who matter less to us who are able to demand our attention.

The great thing about living in the future is that we're constantly connected. And the bad thing is that we're constantly connected. By making ourselves more accessible to the web, we've also made the web more accessible to us. Today there are dozens of ways of letting anyone with a smartphone know you tried to reach them. Most of the time this is convenient and empowering. But sometimes it's also troubling and stressful. When your phone is constantly buzzing with notifications of all the e-mails, IMs, texts, Pokes, reshares, reblogs, and retweets from your entire social circle, it can be hard to focus on the people who are most important.

By putting ourselves online, we can be in touch with more people than ever before. But all that connectivity presents giant contradictions. Today, we live in a world where you can donate a kidney to somebody you've never met, through Facebook, but you can also sit next to someone at work for years and barely ever talk to them, communicating entirely through e-mail and instant message. You can keep in touch with every single person you went to high school with but then go home and spend "quality" time with your family with everyone glued to their laptops, tablets, and smartphones, entirely ignoring one another. We live in a world where you can Skype with someone thousands of miles away at the push of a button but you may need a dozen back-and-forth e-mails to schedule time to see a close friend.

And even though I love my phone and tablet, our shiny, beeping gadgets are now competing with our actual loved ones for attention.

Occasionally I'll be in bed at night, trying to stealthily answer a few final e-mails on my phone, burying it under the blankets. Unfortunately, a bright, glowing rectangle clearly visible through the covers is a little hard to miss.

"Randi, getting a bit of Candy Crush in?" my husband whispers into the dark.

Guess I'm not very stealthy.

Our phones are demanding more and more of our time, and we're giving in. But we're not entirely to blame for this phenomenon. Studies have shown that checking your smartphone can be as addictive as using drugs. According to Dr. Peter Whybrow, a psychiatrist who runs the Semel Institute for Neuroscience and Human Behavior at UCLA, the smartphone is even a kind of "electronic cocaine." Because our brains are wired to seek out novelty as a reward, all those constant updates and notifications you receive from your friends basically give you a little hit every time you tap "refresh." As Dr. Whybrow said, "With technology, novelty is the reward. You essentially become addicted to novelty."

When someone "likes" something you do, your brain receives a little burst of dopamine, a chemical that the brain produces to indicate a reward. That's what makes our gadgets so addictive. Every time we get a notification, we're hoping for another hit.

This is why during dinner, when the phone buzzes, we have to resist the urge to look at the screen, at least until our dining companion goes to the bathroom and is just out of sight. Or we find ourselves

Instagramming our meals instead of just insta-eating the food and liking it the natural way. Sometimes it seems that brunch will never be enjoyed again without someone first hashtagging the hash browns.

Distracted dining has become so pervasive that some restaurants are trying to combat incessant mobile phone use by offering separate food photography time slots or providing a list of preapproved sample food photos with the menu.

That's one way to deal with it. But maybe there's a better way.

There's a lot of talk these days about work–life balance, about managing career and family, about how to "have it all." But this discussion is really more about achieving a *tech*–life balance. It doesn't matter what time you leave the office if your head is buried in your computer as soon as you get home. It doesn't matter that you can instantly e-mail people around the world if you haven't had a face-to-face conversation with the ones right next to you in weeks. It's wonderful to have dinner with friends, but are you really in the moment if everyone is texting at the table? If you don't take the time to invest in the "life" part, then finding balance will always be a fiction. You need to have an "all" in order to "have it all."

You need control over your devices instead of letting them control you. Technology is a tool, and whether it creates order or chaos in your life depends on how you use it. The technology itself is neutral. It's up to you to use it in a way that *enhances* your life and doesn't detract from it.

As technology becomes more advanced, the effect of technology on our personal lives grows every day, and the need to solve these challenges becomes ever more urgent. Smartphones are already challenging our ability to live in the moment, but soon we'll have even more incentives to retreat into our digital cocoons.

Google Glass is the latest sensation geeks are swooning over. Glass is a pair of spectacles that sits on your head, and one of the lenses projects a small display directly into your eye that can show you anything on the Internet, from YouTube videos to Wikipedia articles to porn. It can also take photos and record videos of anything you're looking at and post them straight to your social networks. Basically, it's a smartphone on your head, with limitless access to information.

It's an amazing innovation, but it also presents major challenges. What does it mean for human relationships if eye contact, the one remaining sign that someone is interested in what you're saying, is no longer an indication that a person is paying attention?

My whole life as a young girl, I dreamed of having a big wedding and celebrating with hundreds of people all around me. But after spending years working at Facebook, documenting every moment of my life on social media and keeping in touch with thousands of people every day, something changed. When it actually came time to plan my wedding, I found that I craved something very different from my original childhood dream. I craved intimacy. I didn't care about that three-hundred-person wedding anymore. I didn't need to use my wedding as an excuse to see people from back in high school and college, because I already knew every detail about those people's lives, thanks to Facebook.

In late May 2008, I traveled to Jamaica and celebrated my marriage to Brent with a small gathering of my closest friends and family. I wanted real quality time with the people who mattered most to me, and I wanted the most rare and precious gift of all from those people: their attention.

That's what I got. And it was the greatest moment of my life. We spent three fun-filled days on the beach with our closest friends and family. I had meaningful conversations with every single one of our guests. I had the undivided attention of all the people I loved and cared about. And the wonderful man of my dreams—the one who had encouraged me to join Facebook and follow my passions, the one who had eventually followed me out to California to support my career, the one who was always there for me no matter what—was now my husband.

Attention as Currency

In this new online world, our attention comes at a premium.

Before the age of mobile devices and insta-connectivity, if someone was talking to you it would have been considered terrible manners to pick up a newspaper and start *reading* in the middle of the conversation, or call up old high school acquaintances to see what their babies were up to. But now, thanks to the smartphone, things like that happen all the time.

The IM dings in the middle of work, the text message buzzes in your pocket during a movie, the phone call comes in while you're driving, and you check Facebook while you're visiting Grandma. You can be talking to someone, anyone, and your attention will inevitably start to drift as soon as the phone beeps or buzzes, or fails to beep or buzz for what seems like too long.

The course of a day can feel like a competition for our attention, or the attention of others. Smartphones and social media have done much for our lives in recent years, but they've also seemingly taken away our ability to be truly *present* in any single moment. It's now become socially acceptable to not give a hundred percent of your attention to whomever you're meeting or speaking with.

A 2013 study by researchers at the Injury Prevention Center at the

University of Washington found that nearly one-third of the 1,102 people they watched cross the street at twenty "high-risk" intersections were also on their phones talking, texting, or listening to music. Remember when your mom told you to look both ways before crossing the street? Now you also have to remember to look up.

Similarly, you could be *physically* at work, but if you're checking your social media accounts all day, you're not giving your employers what they're really paying you for: your full attention. (Unless of course, you work for Facebook. In which case, it's awkward if you're not on Facebook all day.)

Your presence is no longer a sign that you're actually paying attention. So, "attention" is something that today is more important than presence or location. In fact, attention is so valuable that it is a kind of social currency. When a random acquaintance sends you a dozen long-winded e-mails asking for an unnecessary phone call, that person is making a massive withdrawal from your attention bank. When you spend quality time with a friend while you're not being distracted by your phone or e-mail, you're investing attention in your relationship and making a big deposit into that person's account. Your attention is a limited and highly valuable resource that you should spend how *you* feel is most important, whether with your friends, your family, your work, or yourself.

Of course, none of this should be understood as a way of *literally* monetizing what are essentially human emotions. You shouldn't look at your kids and say, "No time to talk. Don't you know how *valuable* Mommy's time is?" That would probably give them a weird complex. You can't exchange units of attention like currency, and in a sense, even though some people may seem to not be "worth" your time, that doesn't mean you can ignore everyone who isn't "time-valuable," since that sort of makes you a jerk.

Paying attention to some people and not others doesn't mean you're being dismissive or snooty. It just reflects a hard fact: there are limits on the number of people we can possibly pay attention to or cultivate a relationship with. Some scientists even believe that the number of people with whom we can maintain stable social relationships might be limited naturally by our brains. Robin Dunbar, a famous professor of evolutionary anthropology, has theorized that our minds are only really capable of forming meaningful relationships with a maximum of about a hundred and fifty people, a figure known today as "Dunbar's number." Whether that's true or not, it's safe to assume that we can't be real friends with everyone.

One of the strangest things about human psychology is that sometimes we prioritize our relationships in strange ways, and that affects how we allocate our attention. Think about this fact for a moment: people feel the most pressure to pay back their financial debts to the people they are least close to, and conversely, we feel most comfortable letting debts slide with people we are close to and know we'll see again in the future. Psychologists were able to correctly predict the nature of a friendship by observing the speed of repayment.

Perhaps this helps explain why we feel the most pressure to respond to the e-mails of people we don't really know, and we let letters from our relatives fester in our in-boxes. When a work colleague or casual acquaintance writes an e-mail that we ignore, we feel the pressure to pay back the attention debt we incurred when that person sent the e-mail. But if a friend lends an e-mail's worth of time, we may not feel that same pressure to return the favor so quickly, because we know we'll have the opportunity to connect with them in person soon.

While we were working together at Facebook, my brother Mark would always respond whenever his girlfriend, Priscilla (now his wife), called him. No matter what he was doing, Mark prioritized his attention to her over anyone else. Brent and I have taken a similar approach, and no matter how busy I am, if I see "Brentie" come up on my phone, I do my best to answer it. If my attention is the most precious thing I can give someone, then surely nobody is more worthy of my attention than my family.

Technological improvements have expanded the ways people can both give and receive attention, but when too many demands are placed on our attention, it can make life difficult and strain our personal relationships with those who are most important to us.

When deciding whom to pay attention to, we now need to understand the difference between our friends and our "friends." It's hard to remember now, but there was a time when "friend" was still only a noun and not yet a verb. Back then, our friends were the people closest to us, whom we hung out with regularly, drank with, and confided in. The era of Facebook and social networking changed all that. Now, a "friend" can include anyone from a best friend to a secret nemesis, a work colleague, a distant relative, a neighbor's dog, and Kim Kardashian.

Not only that but different methods of communicating make varying demands on our attention. If someone, somehow, manages to handwrite a letter or send a gift, we have a few days to acknowledge it. A phone call is less demanding than a video chat, because we can do other things and don't have to constantly readjust our hair during the call. An in-person chat with our cell phone turned off is the highest level of attention we can pay to someone.

Thinking of our limited supply of daily attention as a kind of currency may help us prioritize our responses to people. We feel a lot of pressure to respond to all the e-mails in our in-boxes, to get to that mythical unicorn of "in-box zero," because otherwise, it feels as if we have debts hanging over our heads. There are even services that take the conceptual value of attention quite literally and charge the senders of commercial e-mail a premium. This is a pretty big shift for technology companies. In the early days of social networking, every service tried to encourage a social land grab and amass as many friends as possible in a kind of giant popularity contest. As people are now becoming oversaturated with online relationships, the pendulum is swinging back and new services are starting to cater to those who would rather be connected to fewer people whom they truly know and care about. One example is Path, the social network that limits your maximum number of friends to 150, and Couple, an app for, well, couples.

In the end, we have to focus on the people who matter most to us.

This year for my mom's birthday, we went away to a spa together for two days. Over dinner, we toasted to a great trip, but most important, we toasted our promise to one another that even in this age of constant distraction and ever-busier lives, we would always find the time to celebrate and give each other our undivided attention. It's a promise I hope to keep for many years to come.

Going Out Alone—with Friends

It's clear that we need to seek out tech–life balance and be mindful of where we're paying attention, but what, exactly, does that mean? How do we do this? What can we do to live more in the moment?

There's no easy answer, but I think it really begins with a shift in mind-set. We need to find a meaningful way to integrate the new technological advancements into the fabric of our lives.

We're not going backward. Tech is here and advancing. We can't detach modern friendships and relationships from technology. Sure, occasionally you still meet a few technophobic Facebook refuseniks, who refuse to participate in social media, text messaging, or online photo sharing. Of course, we need to be respectful of everyone's preferences and varying comfort with tech, but there's no doubt that it becomes harder and takes more effort to keep in touch with those people. We'll likely only do so if they are very important to us.

New technology has revolutionized our ability to connect and reconnect with our friends and frenemies, our loved ones and ex-loved ones, no matter their location or ours.

Once upon a time, the messages we wrote in our friends' high school yearbooks were likely the last words we would ever have with them. This made yearbook signing a very momentous and solemn occasion. I would take up full pages in the tiniest writing I could scrawl for my best friends' yearbooks, and confessions were left in the margins. But today, we can stay in contact with our elementary or even preschool friends, if we can find them.

The most beautiful moments of our lives can be captured and shared as they happen. Events can be organized on the fly, and new friends can easily be introduced to old ones. And baby photos—we can never have too many baby photos. (Well, okay. I'm biased.)

Of course, as with any technological change, some things can go terribly wrong. It's easy to spend hours online ignoring your loved ones or feeling lonely. Birthdays are now stressful and complicated: Is it okay to just post on Facebook? Do I need to send a text message? A card? A phone call? Is it okay to not acknowledge a birthday if we're really only acquaintances? What if I forget someone's birthday entirely? And on my own birthday, *how on earth am I supposed to respond to all those messages and well-wishes?* Oh boy.

So, what's the answer? It's all about reminding ourselves to live our online lives in moderation and enjoy tech-free moments with those close to us. If attention, scarce as ever, is a sort of currency today, then we might as well spend it cultivating meaningful experiences in our lives and with our friends online as we would offline.

This is not to say that we should spend long hours deeply contemplating every interaction. Sometimes tech just gives us brief, fun, and transient moments, like when we scroll through an Instagram feed, retweet a funny tweet, or forward weird YouTube clips to friends for no other reason than to make them laugh or enjoy a moment of downtime for ourselves.

We need to keep an eye on our daily balance of attention. Cat videos shouldn't edge out the attention we pay to our actual cats. If we're at a concert, we shouldn't spend more time seeing it through the tiny screen on our phones than with our own eyes.

If you're at the Grand Canyon, and before you lies all the great and encompassing majesty of the natural world, stop uploading #canyongrams every few minutes. Spend time appreciating the beauty of the moment, and only upload the image that evokes the feeling you want to bottle and return to for years to come. Just because you *can* document your every waking moment doesn't mean you *should*.

This is the same reason that "going out alone" is a thing now. No, it doesn't mean dining solo. It refers to a "retro" trend in which young people meet up with their friends but leave their phones at home, as a way of getting closer to their friends by giving them their complete and undivided attention.

I recently benefited from this trend by accident, when a colleague said to me, during a recent lunch, "My phone just died, so you're going to have to deal with my complete, undivided attention for the next hour." That "hour" lasted until we found a power outlet. But for that brief moment, it was pretty great.

Some people are taking the de-smartphoning trend even further and taking a "digital Sabbath" on Saturday or Sunday. If we want to make tuned-in time with our families important, this may be a nice thing to try. Although, good luck prying that smartphone away from a teenager. Perhaps make a point to "go out alone" as a family for dinner one night, and leave everyone's phone at home before getting into the car.

Most of the complexities, demands, and awkwardness of modern friendship online can be traced back to the problems of a poor tech–life balance. When people build up expectations of their friends' actions offline or online—and those expectations aren't met—that's when disagreements, resentments, and hurt feelings start.

If it's been a while since you've seen a friend you usually text, maybe it's time for some actual face time. If Instagram is the modern equivalent of sending a postcard—well, you wouldn't spend your entire vacation at the beach writing postcards, would you?

For people you're really close to, a birthday wall post that reads "Happy birthday!!!" isn't going to cut it, even if you use three exclamation points. Pick up the phone and make a call. Let them know you care.

Sure, remembering somebody's birthday once meant *something*, because you had to make a conscious effort to do so. In a world without Facebook, it would be *creepy* if Josh, who sat next to you in eleventh grade advanced-placement English class, wrote you a letter on your birthday to just say happy birthday.

"Dear Randi," it might begin. "I know we haven't spoken since 1997. I hope this letter finds you well. As it happens, I have remembered that today is your birthday, and so I'm writing to wish you a very happy thirty-first! Party, party, party time!! Enclosed, please find a photo of my dog and three pictures of sunsets. Do you like them? If so, write me and let me know that you do, at the address listed above. Very truly yours, Josh."

At that point you would need to change your name and move to a new address, which is not convenient at all.

Things are different now. It's perfectly normal, acceptable, and even nice for the Joshes of your past to wish you a happy birthday. Today, personal information such as your birth date is readily accessible public information and the sort of thing you would probably share as part of your authentic online self.

This is why a proper tech–life balance, which sets clear expectations regarding the new online boundaries of behavior, is crucial to avoiding hurt feelings. Social networking has made it so incredibly easy for anyone we've ever met to wish us a happy birthday that the greeting is inadvertently a less valuable indicator of closeness than it once was.

When a big life event occurs—a marriage, a move, a new job, or a new significant other—it may be a good idea to give close friends a pre-Facebook view of these major life changes. It will make them feel closer to you, and avoid potential offense, than being lumped into the mass of people on your friends list would.

Brent proposed to me on a beautiful Valentine's Day evening in 2007 at the Ritz Carlton in Half Moon Bay, California. After the initial enthusiasm, tears, surprise, and shock had set in, I decided it was time to make some phone calls. We quickly made a mental list of everyone in our lives who would be supremely offended if they found out about our engagement via only a Facebook post and not directly from us. Luckily, Brent had been super sneaky and had already given a heads-up to my family (apparently my mom was at home just waiting for the confirmation call), our close friends (many of whom had helped him pick out the ring and plan the proposal), and my team at work (so he could whisk me away midweek for an overnight trip to Half Moon Bay), so the list of people we had to call was pretty short.

The next step was to post the "ring shot" on Facebook. If you've ever spent ten minutes on Facebook, you'll know exactly what I'm talking about.

But then I put the phone away for the rest of the night and didn't check it again until late the following day. I wanted us to really enjoy our time together and start off this new chapter in our lives by giving each other our undivided attention.

In the early days of Facebook, when everybody in the company was not only a coworker but also an online "friend" of everybody else, a few of us went on a fun-filled trip to Napa. Upon returning, I was confronted by a colleague who had seen the posts about our trip and demanded to know, through stifled tears, why she hadn't been asked to come along. I didn't have a good answer. Nobody had excluded her; it was just that this excursion was thrown together at the last minute. I didn't mean to hurt her feelings. I wasn't even in charge of the guest list. But she felt sensitive about being left out, and seeing all the photos of her friends having fun without her made her feel jealous and unhappy. Sometimes the very best aspects of being connected are also the aspects that are hardest to stomach.

This new phenomenon, which has accompanied the rise of social media, has an acronym: FOMO, which stands for "fear of missing out." FOMO refers to the feelings of jealousy and inadequacy experienced upon seeing the impossibly awesome lives of your friends, and studies have shown that this is a real thing. According to a study by Dr. Andy Przybylski, published in the July 2013 issue of the journal *Computers in Human Behavior,* people who recorded the highest levels of FOMO had the lowest levels of satisfaction in life.

The conclusion was that FOMO could actually "distract you from making the most of your time in the here and now."

Is there a cure for FOMO? We need to remember that we're seeing only a tiny snippet of what a person chooses to post. What people post or don't post on social networks isn't indicative of what they're actually doing. We have no idea what someone's life is like behind closed doors. We're not seeing photos of the dog peeing on the rug or of people fighting with their significant others.

Probably other people are looking at our posts and feeling the exact same way about us. Yet we allow ourselves to feel jealous and competitive. Sometimes when I see other people having fun, celebrating success, and (humble)bragging about it online, I get familiar pangs of envy, especially if it's been a tough day at the office or I'm home with a sick child. But then I remind myself that it's always better to feel happy for others. Tomorrow, the situation could be reversed, and you would want your friends to be happy for you! If people post something that truly makes you feel left out or upset, call them up and *talk it through*. They're your friends for a reason. Give them the benefit of the doubt.

Recently, there's been a trend of people sending out actual "non-invites" to events. These are apologies to people you know you're not going to invite to, for example, your wedding, and you want to give them the courtesy of a heads-up, presumably to defuse the awkwardness of them seeing the photos afterward online and feeling surprised. I personally think this is a terrible thing to do. I understand that it's difficult to tell people they aren't invited to your event. It's never fun to confront people, or surprise or disappoint them. But this is exactly the kind of situation where it's better to address things directly. If people feel close enough to you to be really taken aback by not receiving an invitation, you need to address that in person or over the phone rather than hide behind a computer screen or a post-age stamp.

There is one additional, extremely tricky part of online relationships: how to end one online. Even talking about unfriending is uncomfortable for many people. Young girls in the Girl Scouts used to sing "Make new friends and keep the old," not "Make new friends and unfriend the old." And there's nothing quite as uncomfortable as running into someone in person who you've unfriended. A 2013 research study by Chris Sibona at the University of Colorado Denver Business School found that 40 percent of the 582 respondents admitted to purposely avoiding, in real life, people they had unfriended on Facebook, especially when the reasons for the unfriending were bad.

During the writing of this book, I had several animated discussions with friends about the pros, cons, and etiquette of removing people from your friends list. Most unfrienders fell into one of two camps. There are those who never unfriend anyone because it's mean, and they don't want to lose a potentially valuable contact, colleague, or friend of a friend. A subset of this camp consists of those who are so Zen that they just don't care who's on their friends list. Then there are the enthusiastic, serial unfrienders, deleting people for even the slightest offense or as soon as a friendship has outlived its promise.

I am a proponent of a middle way. Unfriending should take the form of a periodic spring cleaning of people you may have met at some point but with whom you haven't had any meaningful communication or interaction in a long time.

I know. Unfriending feels mean, like you're going out of your way to erase a person from your life, and you owe them an explanation. But the act of occasionally pruning mystery "friends" from your friends list is a perfectly okay thing to do and ought to mean nothing more than saying, "Dear Acquaintance, I'm sorry, but I've forgotten who you are and I'm not sure I want you seeing photos of my kids."

When you do unfriend, do it quietly and discreetly. Don't be a jerk about it, and don't post something to your wall about how *lucky* your remaining friends are for making the cut. Nobody likes to be unfriended, so don't brag about it.

As for people whom you do interact with or see in person but are currently having an argument with, unfriending *them* is an entirely different matter. If you can work things out, then there's no reason to use the "nuclear option" of immediate unfriending, leaving them standing alone at the party, so to speak. Maybe it's better to let the friendship lie low for a while and unfriend during a regular spring cleaning session in the future.

Finally, if your relationship is torn beyond repair and there's no hope of reconciliation, or if a person is toxic, harmful, and beyond help, then unfriend. But do so solemnly, for you might also be unfriended one day.

In the end, the new rules of the digital world are like the old rules: they center on empathy, understanding, and common sense. Always put yourself in other people's shoes, care about the real people on the other side of the screen, and most important, always make the effort to invest time and attention in the people you care about.

Stop and Smell the Flowers

When I was in tenth grade, I had a music teacher who made a big impact on my life. One of his most memorable habits was to tell his students to "stop and smell the flowers." We were fifteen or sixteen years old at the time, so we used to laugh at this and dismiss it. "Have you stopped to smell the flowers today?" we'd joke before entering his class.

A couple of years ago, he passed away. As I thought about the role he had played in my life, I found myself fondly remembering that enigmatic phrase: "Remember, kids. Stop and smell the flowers." Why did he used to say it? What exactly did he mean?

Finally, while writing this book, I completely understood what he meant.

In the spring of 2013, I spent a few weeks in Tokyo with my husband and son. When I wasn't writing, I made a conscious effort to leave my phone behind, unplug, and spend time paying attention to just my husband and son—no phone, no tablet, no computer. I was completely unreachable to the outside world for hours at a time. Japan is a beautiful and mesmerizing country. To savor every moment, I had to *live* in each moment. And even though Asher would probably not remember the trip, I wanted our time and experiences together to leave an impression.

In the evenings, happy and filled with the feeling of a day well spent and every meeting, visit, and encounter stuffed with productive moments and honest face-to-face conversations, I would open my laptop and dive back into the online world. The browser would instantly fill with tabs for e-mail and various websites.

You'd think it would be a stressful return to the reality of a hyperconnected existence. But honestly, it wasn't. I was refreshed and better able to prioritize what I wanted to respond to—what needed an immediate response versus what could wait.

Tips for Achieving Tech—Life Balance in Your Personal Relationships

Tech Should Make You Closer to Friends and Closer to Friendship

It's amazing to have the opportunity to be instantly connected to thousands of people. But your attention is valuable. Make sure all that connectivity doesn't come at the price of your attention to your closest friends and loved ones.

Break Your Digital Addiction

It's so tempting. You hear the buzz. You can almost taste the message waiting for you. But if you're in the middle of a conversation, an activity, or driving, don't check it. The world will not end if it takes you fifteen minutes to respond. And you'll be amazed at what you gain from your interactions when you're truly in the moment.

And If You're Really in Need of a Digital Intervention . . .

Consider introducing a digital Sabbath into your week. Whether it's a whole day without your devices or simply thirty tech-free minutes here and there, give yourself the chance to detox, unplug, and just enjoy the world and people around you.

Birthdays and Unfriending

You probably didn't know that the most popular day to get unfriended is on one's birthday! You see someone's name pop up and if you haven't connected in a while, that person just might not make it to another birthday as your "friend." If we're lucky enough to keep having birthdays and to keep meeting so many people that we have to occasionally houseclean a few, those are high-class problems, my friend.

For most of us these days, in our culture of responsiveness, work is often an unwelcome and uninvited guest on our vacations. According to a recent study by Fierce Inc., 58 percent of workers feel absolutely no stress relief from their vacations, and 28 percent return even more stressed than they were before they left. Boston Consulting Group (BCG) found that employees who disconnected from work for at least one full weekend day each week reported higher job satisfaction and had an increased likelihood of staying with the company longer.

According to some experts, checking e-mail only a few times a day, not checking e-mail first thing in the morning, and limiting the amount of time spent answering e-mail altogether, are highly effective productivity tools. Researchers at the University of North Carolina have found that when people take time out to meditate, they feel more upbeat and socially connected, and they begin to change their social habits. I hadn't seen it in action in my own life until this recent trip.

Of course, I'm not saying that everyone can or should ditch their e-mail and phone for hours at a time. Abstaining from tech completely is generally not a good way to stay employed or run your social life.

Sometimes, however, if we want to really live in the moment with the people we're with, then we have to turn off the phone and disconnect. We own the devices, not the other way around. There is a time to use technology and a time to put it away.

When it comes to finding tech–life balance, life always comes first. Don't forget to stop and smell the flowers.

chapter 6

 **LOVE**

Love in the Age of Facebook

It was a cold winter night in 2001. I was in Boston at a Harvard rager with my a cappella crew, the Harvard Opportunes.

Let me clarify that by "rager" I mean it was a relatively fun party. It wasn't quite as intense as a scene from *The Social Network,* but unlike most of those scenes, this one actually happened. There was, however, an ice luge. Down which only the highest quality Wolfschmidt vodka was flowing.

Somewhere around the time the cheap vodka was beginning to taste somewhat tolerable, I met a lovely guy with a beautiful South African accent named Brent. The fact that he spent any time at all chatting with me and the Opportunes meant he must have really been a stand-up guy. (Let's just say that the more we drank, the more we sang. . . .) But we talked for a while that evening. And at the end of the night, we exchanged IM screen names. In those pre-Facebook times, that was a big deal.

Of course, there were a couple of minor annoyances. For one, he had a girlfriend. Also, on the way out of the party, I made a graceful exit by falling into a bush. Oops.

The next day, I added "BrentT" to my AIM buddy list. I figured I'd never hear from him again.

Fast-forward two years.

A few months into my tenure with Ogilvy & Mather and the Naked Cowboys, I was working on my desktop computer while simultaneously chatting to my teammates on AIM about some edits to a client project. I was waiting for a colleague to get back to me when I heard a familiar *booeoop* sound.

It was BrentT.

"Hey there, Peggy42st, long time no see. Just wanted to say hi and let you know that I just got back from backpacking around Europe and I'm working in New York City now. If you're free, maybe we can hang out sometime?"

Intriguing.

We arranged to meet later that night for margaritas at the bar below my office. I was anxious for the workday to end, and of course, as I was getting ready to leave, my boss tossed me a few more menial tasks to complete. Ten minutes late turned into twenty minutes, which turned into forty-five. Because we had arranged everything over IM, I didn't have Brent's cell phone number, but when I turned up at the bar over an hour late, he was still there, looking cool, with a salt-free margarita ready for me. We caught up and talked about work and life. He had broken up with his college girlfriend. We were both single.

By the second margarita, I had fallen hard for him. Twice, if you count the bush incident.

A few months later, my brother started "The Facebook." By then, Brent and I were in a relationship—a relationship that blossomed into love, and then into marriage and family. So, I managed to bypass many of the experiences and complexities of dating in the digital age. Despite that, I still have a profound appreciation for the role technology played in making our relationship work.

If we hadn't swapped IM names on that fateful evening, we would not have been able to serendipitously reconnect later. At the time, that was tech-enabled social life/dating.

Technology has completely changed all aspects of dating and romantic relationships. Countless apps and websites help people find potential mates to consider. Texting, video chatting, and social networks have created a whole new set of rules for initial courting and the early stages of relationships. Online sharing and a person's relationship status mean that when you take a relationship to a new level offline, you also do so online. And ending a relationship means breaking up with that person in the digital world as well.

Once, if you were interested in someone and wanted to get to know him or her, you had to single-handedly build your own profile of all that person's likes, dislikes, friendships, and interests in slow motion. You would try to run into the person at parties or discreetly ask your mutual friends to arrange "chance" encounters. Now, from the moment you see the words "friend request accepted," you can have a complete dossier on the person's life in seconds and find out where that person went to school, who that person's friends are, and what that person wore for Halloween last year.

Before, if you met someone at a party and wanted to know if he or she was dating someone, you'd have to do some potentially embarrassing real-world reconnaissance. That information is now freely and easily available, along with—if you're really too late— photographs of the person's spouse. Now, when you start dating someone from outside your social circle, a Facebook friendship can establish that you share mutual friends, tastes in books, music, movies, and a sickeningly cute dog named Boo.

Lovers, separated by distance, were once forced to rely on letters and timely phone calls to connect. Now, the Internet makes communication easy across any distance.

Of course, things can go terribly wrong. Social networking can make everything go from weird to bad, fast.

The Broadcast Relationship

When Facebook started, it sort of resembled its namesake—an actual book with faces. Everyone's home page was an identical static page that displayed new wall posts and messages but little else. If your friends edited their profiles, the only way for you to find out about it was if you actively visited their profiles and worked out what had changed since your last visit. It was a tedious game of spot-the-difference. Social networking was about to change the world, but in 2005, Facebook was a pretty basic reference guide.

Reasoning that the online social world would be vastly improved if users had instant, easy access to the social lives of their friends, Facebook created News Feed to automatically broadcast users' posts, profile changes, photos, and friendships and collate that information into a single list of stories that everyone would see when signing in. Today, News Feed is an integral part of every user's Facebook experience and what makes the service so incredibly powerful. Instead of just a home page, Facebook now feels like an actual home—a place for people.

A few weeks before News Feed launched on the site for everyone to use, Facebook employees got to test it out. This is what is known in Silicon Valley as "dogfooding"—people who work at tech companies "eat their own dog food" by using their own products.

Cue one of the very first awkward Facebook relationship updates.

Two Facebook employees—let's call them Joe and Sara—were dating at the time. They had a real hot-and-cold relationship, and everyone was aware that things had been a bit frosty lately. I was close friends with Sara—once a week we'd go to the gym after work and then catch up over salads and white wine—so I witnessed most of the drama firsthand. But we all just assumed they'd get back together, as always. So, you can imagine everyone's shock when the entire office found out—via News Feed, the very first morning we started testing the product—that this time Joe and Sara had broken up for good.

That day, as people throughout the office logged in to Facebook, they were greeted with the top story in their News Feeds, a bright red broken heart next to Joe's and Sara's names. "Joe and Sara are no longer in a relationship," read the short and brutal caption.

There were murmurs around the office, gossip in the bathrooms, and sympathetic and curious glances toward the new singles, who had likely figured that they would lie low for a few days before word started to spread.

But, hey, at least News Feed worked.

We had stumbled onto something incredibly powerful, and it wasn't because we had succeeded in making interoffice dating even more awkward than it was already. We had changed something fundamental about the nature of social interaction. No longer would it be a challenge to stay up to date with your friends' lives, and no longer would you need to hang out by the watercooler to hear the latest gossip. That information would be broadcast to you directly, like a daily friendship gazette featuring just you and your friends. For the first time, technology was making news about friends as newsworthy and accessible as news from a newspaper.

Now, don't get me wrong. The Joe and Sara story highlights an awkward moment, but the vast majority of the updates we see are incredibly positive. Even if we could go back to the world before the feed, we would lose so much by doing so. There's a reason we used to hang out at the watercooler to gossip! Now the watercooler is online, and all our friends are there. But as Joe, Sara, and the rest of the world soon figured out, this means our romantic relationships have an unprecedented new public dimension that has yet to be fully understood.

This is what I call the "broadcast relationship."

The broadcast relationship isn't a phenomenon limited to News Feed or even Facebook. It's not something that we choose to have or that we can avoid. It's the reality of what all our relationships have become, thanks to being connected in so many ways through the Internet, social networks, and mobile phones. Now that we can share and experience every moment of our lives with our friends as soon as they happen, technology has turned relationships into entire narratives that now play out online like miniature soap operas— from first meeting to last, photo tagging to de-tagging, friending to unfriending to refriending.

The result of all these interactions taking place in sight of our friends and the entire world is that relationships are not just defined by the moments experienced together in person but also by the broadcast and the reception.

Think about the presidential debates we watch every four years. A candidate might win the room, but the real peanut gallery is at the other end of the screen, and that's where the debate is won.

We're all candidates running for romantic office. We all want to impress the people we're dating, by taking them to fancy restaurants, springing for the extra-fancy bottles of wine, and putting up with their annoying friends. But now we can't just impress in the moment. The perceptions they form about us are also shaped by how the narrative of our relationships plays out online. What if you don't look good in your photos together? What if your friends aren't excited when you become "Facebook official"? What if you secretly loathe that the other person listed Nickelback as a music interest?

These questions might seem silly. Who cares what your friends think about your date? Why should you care about your online relationship when the real relationship is good?

These questions are important, because in a world of authentic online identity, there is increasingly little difference between our real and our online selves. The two cannot be thought of separately. So, in the world of the broadcast relationship, everything that happens in our relationships affects our online identities as individuals and, as couples, becomes part of our *shared* identities.

This creates all sorts of complex new considerations for relationships. For example, when you start dating someone, pretty much the first thing you want to know is "What's that person's profile like?" That's because you want to know his or her identity and find out if it's compatible with yours. The discussion that always began with "Are we a couple?" now has to end with "When do we make this official online?" The moment you make things official, you're inviting the approval and judgment of your entire online community.

It's not just the things that are announced by consent in a relationship that hit the Internet and that become public knowledge. Everything from the location of a date to the contents of a dinner plate can be tweeted, Instagrammed, uploaded, liked, and commented on, with or without your knowledge and approval.

I've known plenty of people who sneak off to the bathroom during a date to update their friends via Facebook on the progress of it. One episode of the Bravo show I produced, *Start-Ups: Silicon Valley,* featured larger-than-life entrepreneur Sarah Austin "life-casting" her date and getting caught by him in the act. *Big no-no.*

Another friend uses codes on Twitter to live-update her followers on how her dates are going. Only a few of us know her secret code system, and I won't give it up.

It's worth mentioning that, thankfully, most people *don't* live-update their dates. And many people avoid social media altogether during the courting process.

Beyond even these novel pressures on our relationships, broadcasting our relationships can simply be stressful and exhausting, and something that makes the pursuit of our tech–life balance even more difficult. There are three important things we need to think about: intimacy, shared expectations, and identity.

You Can't Tag Intimacy

In early 2008, Brent and I were preparing to go to Jamaica for our wedding. Our bags were almost packed. An airplane ticket had been purchased for my dress (not joking, although I wish I was), and I was due to head down to Jamaica a week early with my mom, to make sure everything was set. Of course, nothing in life ever goes as planned. Two days before I was to leave, we got a call from the rabbi who was going to perform the ceremony in Jamaica, making sure we had been legally married in the United States. Brent and I looked at each other. "What? How could we have missed that oh-so-little-detail of, you know, *getting married before we got married*!?" We weren't sure if he had forgotten to tell us that we needed to first do a legal ceremony in the States, as the Jamaican wedding wouldn't be legal, or if we were the ones who had forgotten, but either way, we were scrambling.

Luckily for us, our good friends Chris and Jennifer knew a local judge who agreed to legally marry us the following day just hours before my plane left for Jamaica. It was going to be difficult—I still had to finish packing, and I had a million things to wrap up at the office before I left the country, and work, for three weeks.

The following morning, I was at the office, racing around in last-minute meetings and tying up loose ends. I was right in the thick of everything to do with the presidential election, and I wanted to make sure it was all left in good shape.

Suddenly, my Outlook calendar flashed. A bell rang from my tinny computer speakers. And up popped a notification: "Invitation from Brent Tworetzky—GET MARRIED."

I stared at it, dumbfounded for a moment. And then I burst out laughing at the sheer absurdity of seeing something so profound appear in such an underwhelming way. Even though it made perfect sense for Brent to send me a calendar invite for our visit to the judge, I had great fun teasing him later about it.

So, technology isn't always best suited for the more romantic moments in our lives. And sometimes Facebook just doesn't cut it. A recent study by Dr. Sam Roberts at the University of Chester, in England, found that people were happier and laughed 50 percent more when they interacted with their friends in real life, versus on social networks. That's a lot of jokes to be missing out on.

Brent and I, like most couples we know, can be guilty of technically spending time with each other but really spending time with the Internet. Evenings that we plan together at home often turn into evenings where we sit side by side on the sofa, both of us on our laptops, not actually speaking to each other at all. If relaxing from a busy day at work used to be something you did socially, it's now a mostly solitary activity, consumed by mobile games, streaming videos, or online shopping—activities where you can physically be in the company of someone else, but mentally you're all alone.

For our wedding in 2008, I had cake toppers that poked fun at how busy and focused we were in our careers at the time: a bride and a groom, both on their BlackBerry devices, looking in opposite directions. Brent said the cake toppers made him a bit sad, but I loved them, and eventually he decided they could stay. Five years later, our cake topper has become something of the status quo for relationships.

I hear a lot of couples talk about the iPad-in-the-bedroom conundrum. This is where a couple gets into bed at the end of a really long, exhausting day, and instead of actually talking or, ahem, *interacting*, both people just take out their tablets or phones and start surfing the web. In 2006, a study of 523 Italian couples by a team of psychologists found that couples who had a TV in their bedrooms had, on average, half as much sex every month as those who didn't. Well, having a phone or tablet in the bedroom is the same thing, except with a few million more channels.

In a recent study by Bayer HealthCare Pharmaceuticals, 28 percent of women claimed that e-mail and the Internet were disrupting their love lives, with mobile devices particularly to blame.

Fingers crossed that the human race doesn't end up going extinct just because we can't find the self-restraint to stop playing Bejeweled Blitz. That would officially be the lamest ending ever.

Sharing Is Caring?

We all know "that couple" on Facebook. Constantly posting lovey-dovey photos, incessantly talking about how great their relationship is, writing daily status updates thanking God for the one and only love of their lives. I used to roll my eyes when I saw these posts. But then I got curious and wondered: What's in it for these couples?

What do they have to prove, and who are they proving it to? Is their relationship *actually* as good as the show they're putting on?

A few years ago, one of my college friends proposed to his then-girlfriend in a very public proposal that played out moment by moment through Facebook photos and status updates. Although it was fun to feel like I was part of the moment, I couldn't help but wonder if it was designed more to impress their friends online than it was for the actual couple. A few months later, they called off the wedding. It seems they had put so much attention into putting on a show of intimacy for everyone online that they forgot to actually have intimacy in real life.

Intimacy is more than just a beautiful photo or a well-written post, designed to get likes or retweets. Just because we're putting on a good show doesn't mean we're actually investing in our relationships.

Whenever people talk to me about a situation like this, I urge them to have a discussion with their significant others and set some ground rules. In a truly intimate relationship, whether platonic or romantic, it's important to be able to hang out "alone," without inviting hundreds of people online to join you every single time. It doesn't matter if the whole world knows you are friends, or dating, or attending cool events. The only thing that really matters is that you actually enjoy spending time together when the cameras aren't rolling.

Brent and I have set up private online groups and accounts where we share moments between us about our son that are meaningful and special. These are things that we want to remember and preserve, but we don't feel the need to share them with all our friends or the entire Internet. Instead of broadcasting, we're narrowcasting. We get the benefits of sharing without the pressure of the peanut gallery.

And the simplest thing to do is to just have moments with your partner that you don't document online. If you find yourself on a beach together at sunset, don't invite the Internet into that moment. Even if a pod of baby dolphins surfaces before you, while the surf crashes softly onto the shore, let it go. Put the phone away. The world does not need another hashtagged sunset.

It's wonderful to see couples enjoying beautiful, touching moments and sharing them online. I love seeing my friends happy, and I love seeing expressions of joy and love. But I always hope that same couple is making sure to enjoy an equal number of beautiful, touching moments that they choose to hold special and unique to just the two of them. It's the intimacy you create when no one else is listening that matters most. And, in this age of social media and 24/7 connection, other people seem to be listening and watching an awful lot.

One more dimension of intimacy I want to talk about is what this means for long-distance relationships. As someone who had one, this is a topic I'm quite familiar with.

When I moved to California in 2005 and Brent stayed behind in New York, we talked online all the time on AOL Instant Messenger. We also spoke on the phone a lot. But we didn't call every day, and our IM conversations weren't scheduled events. We definitely communicated every day in one way or another, but sometimes that meant just swapping a quick e-mail over lunch or sharing grainy BlackBerry photos when we were out and about in the evenings.

A lot of this flies in the face of orthodoxy about long-distance relationships, that you ought to have a routine for communication and that the more you share the better. But this comes back to my opening point about the need for both attention and intimacy.

In a long-distance relationship, intimacy is usually the biggest potential casualty of the distance. And just throwing a lot of time or attention at the problem is not a substitute. You can't just communicate more to compensate for the distance; in fact, that's a surefire way to destroy intimacy. Because when we overshare, that's when we make our interactions less meaningful.

A study released in April 2013 by Dr. Bernie Hogan of the Oxford Internet Institute tested the impact of multiple forms of communications on marital relationships. After examining how twelve thousand couples communicated with each other, Dr. Hogan found that those who used a greater number of media channels to communicate reported no increase in marital satisfaction, and in many cases, the stress and pressure of sharing so much actually put a strain on the relationships.

Brent and I went for a balance: we communicated regularly, but we never treated our communications as a replacement for more meaningful conversations and attention. And by giving each other more space and choice when it came to our daily exchanges, it meant that when we did have a proper phone catch-up every few days, we generally had more to share and looked forward to those moments with greater excitement.

Also, every time I heard that screen-door sound on AIM, my heart leaped into my throat. To this day, every time I hear a screen door open, I naturally assume something awesome is about to happen.

Intimacy is more than attention and goes way beyond just putting on a show. Know how to achieve a balance between the two, and know when to stop broadcasting. It sounds like simple advice, but in a time when we can share everything, knowing that we shouldn't share everything may be the hardest advice of all to follow.

Love Means Never Having to De-tag a Photo

Partly because I worked at Facebook for so long, and partly because it's just my nature, I am pretty comfortable living my life publicly.

Brent, not quite as much. We're constantly debating what I'm allowed and not allowed to post, especially when it comes to photos of
our son. I'm definitely the "broadcaster" in our relationship. (Shock-
ing, right?)

What matters, of course, is that we're open and honest in our
discussions with each other, and we work together to find a happy
medium within both our comfort zones. Hyperconnected relationships can be hypercomplicated, and having candid, trusting conversations with your partner seems like a sensible way to avoid conflict.

The most important thing to do here is to understand what your
shared goals are with technology.

This past year I went to three weddings. At one wedding, the
guests were kindly instructed not to post any photos on Facebook.
At another, the couple had ushers go around and collect people's
phones during the ceremony, surprisingly without having to use any
force. And at the third wedding, they not only encouraged people to
post photos, they assigned a hashtag to the ceremony and had signs
up to encourage tweeting.

For the record, I had a great time at all three weddings, though
after #mikeandnancyswedding, I was a little worried about what
would be posted during the honeymoon. These were three very different weddings. Each of the couples clearly set out with a vision in
mind of the role they wanted technology to play in their special day,
and they achieved success in their different ways because of it.

Earlier this year, my team helped create one of the most interactive
weddings ever, when Zuckerberg Media partnered with Condé Nast to
produce *Brides Live Wedding*. Every single aspect of the wedding was
voted on via social media, from the couples who uploaded videos of
why they should win their dream wedding to the cake, the flowers, the
dresses, the theme, the colors, etc. The final wedding was streamed live

online and on Facebook, in professional television broadcast quality, to millions of people. This obviously took a couple who felt comfortable making their special day so public and ceding control of all decisions. But the event was beautiful and demonstrated how the Internet makes it possible for people to feel as if they are truly there experiencing an event, even from their living room halfway around the world.

For couples in the Facebook era, the same kinds of conversations need to take place about some of the more mundane but still tricky conventions of tech etiquette. "When is it okay to take and post a photo of us together?" "Is it okay if we share a profile photo?" "What if we're making out in the photo?"

I'll answer the last one right now: no. You're welcome.

In all seriousness, though, differing opinions between partners about the level of acceptable online affection can cause serious offline tension. You should be open and communicative about what gets posted, whether you're an overly lovey couple with lots of co-tagged photos and daily declarations of love on your wall or you're against online displays of affection on principle.

It's not rocket science, but it's crazy how many couples never talk about this stuff. The more horror stories I hear, though, the more I'm convinced that being able to discuss and agree on a shared vision of online privacy might just be a basic sign of romantic compatibility these days.

Sometimes it's more than just public displays of affection that get shared. Sometimes it's the private ones. In the smartphone era, people aren't just sharing more, they're sharing more of *themselves*. I am, of course, referring to the phenomenon of sexting. Many adults (it's not just the "kids"!) are sending each other pictures of their formerly private parts. "Dickpic" is a commonly used word now. Look it up in the Urban Dictionary if you don't believe me.

Sexting is one of the riskiest online behaviors for couples. Obviously it's not for everyone. But if you're going to do it, you absolutely want to make sure that your partner has shared expectations of what you have planned. And be careful that you don't tweet more than you intend to. If U.S. Congressman Anthony Weiner couldn't keep his namesake off the Internet when he (repeatedly) tweeted a picture of his "junk," then it's clearly pretty easy for anyone to make a huge mistake here. (Or not so huge, as it seems.)

To Thine Own Profile Be True

In the end, it all comes back to identity.

In the era of the broadcast relationship, our partners and potential partners can now get an incredible sense of our identities in an instant, and our identities combine with theirs to create a shared online identity. This dynamic definitely creates new complexities for our relationships. When people examine our online selves today, they can see things they don't like about us, which may immediately have a negative impact on our relationships, and we don't even get the opportunity to explain ourselves first. It's difficult not to judge a book by its cover, and these days the story of one's life has a cover filled with information and photos.

An online identity can also be an amazing way to help us find the people we really fit with and care about. Having a mutual understanding of each other's authentic identities is the best way of finding the people we want to share the intimate moments with in our lives and who have similar expectations, interests, and values. Simply put, when you're authentic—online and offline—that's when you're most likely to find a keeper.

Facebook is incredibly powerful as a service precisely because people use their real names, identities, and interests. From the start, this was one of the biggest competitive advantages over rival social networking sites. It's how people who know you can find you and add you to their friends list, get a sense of your personality, and choose whether to develop greater connections with you online and offline. It's no different for dating. The more you reveal what kind of person you really are, the more likely it is that someone compatible with you will take an interest, which then leads to conversations, plans, and dates.

Online dating is like the ultimate job interview. The emotional and personal stakes are high. It can also be a brutal process. Online dating is an interview where, with the push of a button, you can be instantly judged and compared to an infinite number of other candidates. This dynamic creates an incentive for people to find interesting new ways to differentiate themselves from the competition and to "brand" themselves in the most appealing light.

There's nothing wrong with wanting to differentiate oneself. It's what's needed today. But when people lie about themselves online, that's when problems start. What's the most important lesson for relationships today? Don't try to be someone you're not online.

One of the reasons Facebook works so well to promote authenticity is because it evolved as a community with a shared set of expectations and conventions around authentic identity, and telling the truth was expected and enforced by our friends. But when we don't have those safeguards—and, in fact, everyone is trying to compete at being the most perfect—a vague interest in *Lord of the Rings* becomes a love of medieval folklore, sporadic attempts at boiling pasta means that a person enjoys cooking, and because someone went to the gym a couple of times last month that person is now training for an Ironman.

This is just asking for trouble. People who post inauthentic or inaccurate versions of themselves on dating sites may find themselves either struggling to explain their exaggerations, or on a date with someone trying to do the same. Ideally, if this happens, *neither* of you is really training for a full Ironman. But posting a misleading profile picture is definitely going to be noticed.

In the end, the truth usually comes out. If you do build a relationship with someone that begins with lies or exaggerations, the digital world makes it easier for those things to come back to haunt you. Your online identity leaves a digital trail across the web, on Facebook, LinkedIn, Twitter, Instagram, and blogs. That record just begs to be Googled, studied, and cross-referenced. So, don't lie about your identity online. The truth comes out offline.

Remember that your identity doesn't just belong to you anymore. The photos and posts you personally make are one side of the story, but the company you keep and the types of comments those people write on your posts says a lot about you as well. We are now judged not only by what we say, but also by what other people say about us.

If you're inauthentic about the friends you accumulate online and add people you don't really know, or people you've met only once in passing, this could cause your *actual* friends to be misled by a false sense of closeness when they meet these people, find them online, or see that you are the mutual "friend." By contrast, if you meet someone who tells you about a friend you have in common, either online or offline, be sure to confirm it, just in case.

This requires we think carefully about the friends we let into our lives and allow to post or tag on our behalf. But it also emphasizes the importance of telling the truth at all stages of a relationship, online and offline.

We all have that "friend" who, even though he or she is only a peripheral acquaintance in real life, acts like our best friend online. That person who likes every single photo, favorites every single tweet, comments on everything, and tags you in photos you're not even in because you're "there in spirit."

Tips for Achieving Tech—Life Balance in Your Romantic Relationships

Be Your Best, but Be Yourself

Your online self is an expression and extension of your actual self. To be successful in the online dating world, present yourself in the best possible way, but don't try to be something you're not. It sounds risky, but the risk will be worth it when you actually meet your online crush in person and you can just be yourself. It's exhausting to put on an act, and in this digital age, since it's easy to check for inconsistencies, the truth will eventually come out anyway.

Sharing Is Not a Replacement for Intimacy

Your phone can be a powerful tool for sharing and connecting with your partner. But it can also cause problems. Make sure you share expectations with your partner about how much you feel comfortable sharing online, and make sure you're on the same page. Be smart and banish devices from the bedroom, put the phone away when it's time to eat, and pay attention to that beautiful person glaring at you across the dinner table.

Know When It's Time for a Digital Untangle

In the old days, when you stopped dating someone, you didn't get to keep going through their photos or know every detail of their newly single lives. This should be true online too. If you keep clicking back to your ex's Facebook profile, it can be really unhealthy.

Your ex will be smiling at you from parties you weren't invited to, people you don't know will write on your ex's wall, and updates will show him or her hanging out nearby with a new sweetheart, or looking for one. Don't keep refreshing the heartbreak. When your offline relationship ends, move on from your online one as well.

Intimacy Is Not a Show for Friends

The broadcast relationship helps you express to your friends that your relationships are honest, happy, and meaningful. But overdoing it can rob you and your partner of intimacy, as well as annoy your friends. Intimate, beautiful moments should be kept private. Sometimes, the only person you need to "share" a moment with is right next to you.

Always Confirm Your Friends' Friends Are Legit

Let's say you have Roger as a Facebook friend, but you've met him only once. Roger meets Laura at a party, and Laura is actually a good friend of yours. She has a fun time with Roger, then goes home, looks him up on Facebook, and sees that Roger and you are friends. *Cool,* she thinks. *Roger must be a pretty upstanding guy, then.* It's easy for Laura to feel a false sense of closeness to this guy, because she validated it by an inauthentic friending on your part. Only friend people you know, and whenever someone you've just met is "friends" with someone else you know, be sure to verify the relationship with your mutual connection.

In the end, we are who we are, online and offline. Your identity and your emotions don't end when you go online in search of love. When you begin to think of the online you as a part of the offline you, then knowing how to interact with the people you care about in the digital world becomes perfectly natural and intuitive. And that's when you can concentrate on getting to know new people—and on listening to your heart.

Earlier this year, there was a lot of talk about a term called "cat-fishing," when Notre Dame football player Manti Te'o found out that he had been cat fished—the girl he had been online dating for three years, had supported through cancer, and believed had recently passed away, was actually another guy, who had created a fake online identity. Turns out Manti Te'o had never had a relationship with that woman, because she had never existed.

Authenticity can be a painful thing for the heart to bear.

Breaking Up Is Hard to Do

One more painful consequence of putting your authentic identity online concerns an unfortunate chapter in many relationships: the breakup. Sometimes things don't work out and love isn't meant to be. When that happens, after you're done handing over the other person's toothbrush and taking back your hair dryer, you face a more serious challenge: detangling your digital selves from each other.

Post-breakup is treacherous in this new territory, because when you date someone, you participate with them in a shared online identity. The photos taken together, the check-ins, the lovey-dovey public messages you sent each other on Valentine's Day, the shared friends and connections—all of these are a core part of your relationship.

Many couples try to remain friends online, even after they have ended the relationship offline. It can seem that the online-only relationship is harmless and that the pain will pass. Because of the shared online identity that relationships create, you can't break up in real life and not break up online. And research shows this is an essential part of healing and moving on from the relationship.

Scientists at the Brunel University in the United Kingdom have shown that staying Facebook friends with your ex creates "greater current distress over the breakup, more negative feelings, sexual desire, and longing for the ex-partner, and lower personal growth."

If you break up but stay Facebook friends, you'll have to experience the pain of seeing your ex going to parties with single friends and getting tagged with a new partner, and you will constantly be reminded of the loss of an online identity you once enjoyed together. It also confuses others and raises questions about whether the relationship is truly over. Is that something you're prepared to deal with?

And what about couples who share a blog or a Facebook profile? Recently, divorce lawyers have reported a significant increase in couples arguing over who gets ownership of online assets and social media accounts, counting them as valuable "property" in divorce hearings.

Interestingly enough, a QMI Agency study showed that 45 percent of people would be happy to be contacted by an ex on Facebook, but at the same time they would be furious if their current significant other was in contact with *their* ex.

We see the effects of technology in all our relationships, but in our romantic lives, the effects are more intense, with the potential to help us find intimacy and affection or break our hearts and crush our dreams. Dating and love in the modern world are complicated, to say the least. And using technology to go in search of love can be a risky business.

Knowing how to express the best version of yourself online, broadcast when appropriate, provide real intimacy, and survive the digital untangle will help you find the tech–life balance that continues to be the cornerstone of all successful personal interactions in the digital age.

After all, computers don't get into relationships. People do.

 FAMILY

The Truly Modern Family

In the fall of 2011, just a few months after Asher was born, my friend Hooman asked me to play a bit part in *Olive*, an indie film he was writing and directing. I've always wanted to be in a movie, and this sounded like an especially cool one. It was the first feature film to be entirely shot on a smartphone.

I was only going to appear in one scene, and thus only had to be on the set for a day. It was a big scene, though, and I was to play opposite Academy Award–winning actress Gena Rowlands. That day, I turned up at the studio early, feeling glamorous and excited for my role as Shoe Sales Girl. I had memorized my lines and seen the really cool cell phone video camera they were using to shoot the movie. But when it was time to be fitted in the wardrobe department, I made a horrifying discovery. I realized that I hadn't shaved my legs in a few days. I wasn't glamorous; I was gross.

Sheepishly, I asked one of the set runners if she would go across the street to Walgreens and fetch me a razor.

The runner laughed. "Wow, I've seen a lot of things working on movie sets, but this is a new one."

I started to mumble an apology. "Well, I just had a baby . . ." But she had already run off across the street. No doubt, she would later be regaling her friends with a story about the poor, pathetic new mom she had met on set.

Luckily, the scene went really well. And when I saw the initial edit, I was shocked at how beautiful footage from a smartphone camera can look on the big screen. We've come a long way. You could never tell that behind that smiling Shoe Sales Girl on camera was just an ordinary mom, trying to balance the excitement and challenges of raising a family in the digital world.

How does technology affect family life, children, and parenting? As we have seen, there are both good and bad, complex challenges and awesome opportunities. But up front, let's get something out of the way—a lesson that my embarrassing on-set experience demonstrates: you can't have it all.

You can't pay attention to everyone, and lately it's difficult to even pay attention to yourself. No matter your age, gender, marital status, financial status, or background, we all have to make hard choices and trade-offs. All the people who expect to be able to combine their home, work, and personal lives without sacrifice are just begging to be disappointed.

I follow a simple mantra. *Work. Sleep. Family. Friends. Fitness. Pick three.*

Sometimes when Asher is at the playground playing with his toys in the sandbox, he'll look up and say, "Cookie?" I gently explain to him that he can't have a cookie while he's also playing in the sand. If you're doing one thing, sometimes you just can't do another—you have to choose.

Every morning, staring bleary-eyed into the bathroom mirror, I remember my mantra and think about what to prioritize. The way I see it, work, sleep, family, friends, and fitness are the essential things we need but can't do every day, or at least not well. Each morning, I pick three, depending on what I need to do and want to do. Obviously if it's the workday I don't really get a choice. But I do get to choose how to spend the in-between moments of the day and the mornings and evenings before and after work.

This morning ritual began when I became pregnant with Asher, in the summer of 2010. Over the next nine months, I took more than two dozen work trips for Facebook, several of them international. At three months pregnant, I did a panel discussion with Anderson Cooper in New York City. Then I got on a plane to give a talk at the Foreign Office in London the next day. At four months, I represented Facebook at the Golden Globes, desperately trying to make my bulging belly fit in with the gorgeous, slim women on the red carpet. At five months, I attended the World Economic Forum in Davos, Switzerland, where I battled fatigue by napping in the corner of a booth in between meetings with world leaders. At seven months, I organized a live talk show for Facebook at the South by Southwest conference, where I waddled around Austin, Texas, conducting televised interviews with A-list celebrities and politicians. And at nine months, I organized the Facebook town hall for President Obama, all the while battling nausea, fatigue, and intense leg and back pain from Asher pressing on my sciatic nerve.

Suffice it to say, I had to get really good at prioritizing, and fast. As you can imagine, I often picked work and sleep that year and tried to spend what time remained with Brent. I didn't get to spend much time with friends and definitely didn't prioritize fitness. I gained nearly fifty pounds while pregnant and am still fighting the final five two years later. All in all, that year ended as one of the most fulfilling of my entire career.

I continue to run my life by this mantra, while mixing things up as often as possible. After all, life is flexible, and we can do more than we expect. And the best part is, every morning we get to make our three choices all over again: Work. Sleep. Family. Friends. Fitness.

How can we have fulfilling family lives and make sure our children get the attention and love they deserve? How can we raise our children to be tech savvy but also safe? How do we help our children find their own tech–life balance?

You can't have it all. But technology can help you enjoy and make use of what you do have. Technology, when used mindfully, can be used for more than just solving the problems created by technology itself.

Does technology make family life easy? No. Does it absolve you of the need to give your kids love, attention, and affection? Absolutely not. Is technology something that is automatically good? No way.

Technology is a tool, and to make sure it's used correctly we need to remain engaged and involved in our children's lives and teach them the right skills and habits to stay safe and productive online.

Dot Complicated—1990s Style

When I was a kid, one of my earliest creative projects was designing and launching a family newsletter. At some point during my early years, I liked to imagine myself as a serious journalist. But lacking all the important journalistic resources—a fedora with the word "Press" on it, a vintage camera, and, er, any actual stories to report—I decided to let loose my not inconsiderable talents on our household.

The newsletter was called Half a Dozen because there were six of us in our family. As you can see, I was already a pretty witty kid. And the content wasn't bad either. The newsletter mostly alternated between matter-of-fact announcements of upcoming events and activities (useful) and elaborate "investigative" reports into household activities (ingenious).

"Half-Drunk Orange Juice Left Open in Fridge," read the headline. "Pulp Culprits Still at Large."

Okay, so it wasn't exactly Bob Woodward quality, but my captive audience of at least five other people made me genuinely excited to write. Getting a kid to willingly dedicate time to any writing that isn't homework is also a pretty big deal. Every week I would log on to my dad's office computer, open up Print Shop, and bash out another page of "news," which would then be printed in black and white on our printer and pasted on the refrigerator door. It was an incredibly satisfying feeling, even if I never did get to the bottom of the Great Orange Juice Mystery of 1991.

Looking back, that newsletter might well have been the precursor to Dot Complicated.

If I were growing up today, Half a Dozen would be completely unrecognizable. It would be a hundred times more sophisticated. And there's no reason it couldn't be an actual newspaper of the Zuckerberg clan. Technology allows us to create amazing, professional-looking content quickly and easily. What a few years ago would have required a serious professional with qualifications in art and design to produce we can now make with our own computers in minutes. And instead of just pinning a piece of paper on the refrigerator door, now the family newsletter, as well as constant smaller updates, can be shared with your household, your extended family, and anyone else you want to include, via blogs, groups, or e-mail.

And that's only the start.

What technology has given us—which we often take for granted—is simple access to information. And this provides amazing benefits for families, as well as some challenges.

I'm always amazed that no matter where I go or whom I speak with, no matter if you're a mom in Silicon Valley, Oklahoma, or Tokyo, we're all just people with the same personal questions about tech and our families.

On one hand, people are extremely excited about what technology has helped their children achieve at such a young age. One man told me about how he helped his nine-year-old daughter self-publish a book on Amazon, something he could only have dreamed of doing when he was her age. Another woman told me about her thirteen-year-old daughter who was already designing her own mobile apps. And almost all parents brag about how their kids are the best tech support they have.

Yet people also have a lot of questions and concerns. What about the new such-and-such app that all the kids are using? How can they convince their kids to friend them on Facebook? How can they have productive conversations with their kids about issues like online privacy and cyberbullying?

It's incredibly complicated to raise children in today's modern, wired world. We are the first generation of parents whose children will grow up entirely online, with every single moment of their lives documented, recorded, and stored publicly. This is especially challenging given that we did not grow up with all of this technology and it is far from second nature for many of us.

We're the first generation of parents whose kids think it's perfectly normal to interact with other human beings through computer screens and who assume that every single screen they see has an element of touch to it. And we're the first generation of parents to be grappling with issues around privacy, safety, and anonymity online, which didn't exist just a decade ago.

So, it's up to us to forge the path ahead.

And the Cat's in the Cradle . . .

In 2011, the Joseph Rowntree Foundation found that children who see their parents drunk are twice as likely, as grown-ups, to regularly get drunk themselves. This didn't even need to be something that happened frequently—the study found that those odds were reached even if children saw their parents under the influence on only a few occasions.

This is just one example of how parents can have a much greater impact on their children's habits than they might realize, and that social behavior can be contagious. When kids are growing up, they are powerfully influenced by our words and actions, and whether we want to or not, whether we are conscious of it or not, we will leave an imprint on our children's personalities and development forever.

Recently neuroscientists have begun to shed light on some of the biological and environmental factors that explain why this happens. One potential explanation can be found in the role that "mirror neurons" play in our brains. Neurons are the brain cells that help us communicate, think, feel, and love. Mirror neurons play a fascinating and unique role among these: they help us learn by imitation. When we see someone riding a bicycle or laughing or dancing, we get an intuitive feel for how to do these things ourselves, and then we learn by imitation.

Another way adults affect their children's development is through cultivating a sense of empathy and connection. A 2007 research study by Dr. Ruth Feldman of Bar-Ilan University detailed in the journal article "Mother–Infant Synchrony and the Development of Moral Orientation in Childhood and Adolescence," showed that when we pay attention to our children, we are activating and strengthening our children's long-term capacity for empathy.

Also, the importance of a child hearing a parent's voice has been shown by researchers to play a critical role in the development of intelligence. A study by Betty Hart and Todd R. Risley at the University of Kansas, published in the 1995 book *Meaningful Differences in the Everyday Experience of Young American Children,* showed that the more words a child hears before the age of three, the better that child will do in school.

The relationships and attention children experience while growing up imprint on their minds and personalities. When we share good feelings and emotions with them, or just talk to them, we are preparing them for life.

Of course, it doesn't work if we're too busy checking our e-mail.

How often do you scold your children for using their phones too much and then find yourself at the dinner table, checking your text messages or answering a work e-mail? How often do you tell your kids to get off the Internet and do their chores or their homework, but then you waste time online yourself? How often are you staring down at the screen of your smartphone instead of looking into the eyes of your child?

People ask me all the time what my rules for Asher are when it comes to technology. For how long do I let him play games on the phone? How much screen time does he get on my tablet every day? Does he have his own iPad? But nobody ever asks me what my rules are for myself.

Because Asher is so young, I've found that it's more important for my husband and me to set rules for ourselves to make certain that we're not the ones spending too much time on our phones or laptops. Whenever I catch myself sending a text message or reading a blog post instead of engaging with my son as he's tugging on my leg and holding out a train or puzzle to play with me, I start thinking about the famous Harry Chapin lyric, "And the cat's in the cradle with the silver spoon," where the dad keeps telling his son that soon he'll stop working, soon he'll stop being busy, soon they'll hang out. But by the time he actually gets around to it years later, his son is too busy. I know that one day I'm going to be desperately trying to get Asher's attention, and he's going to brush me aside, half paying attention as he buries his face in his phone.

During a recent family trip, as we used Google Maps on my husband's phone to guide us, I felt weirdly conflicted. On one hand, this technology was essential for navigating. On the other hand, my husband was spending so much time engrossed in his phone I felt like he wasn't seeing the sights around him or experiencing the adventure. I worried that we were setting a bad example for our son.

Many of my friends complain that at Thanksgiving, the adults are way worse than the teenagers when it comes to checking their phones, texting, and answering e-mails at the dinner table. A recent, terrifying article published in USA Today reported that adults text while driving at a higher rate than teenagers. Even though texting has long been thought of as a "teenage" problem, 98 percent of those texting adults admit that they understand how dangerous texting and driving can be. Even scarier, 30 percent of moms text while driving with their babies or young children.

It's really, really hard when we're in the moment to remember that our children model their behavior after us and we need to be the ones setting a good example. When that work e-mail is burning a hole in our pockets, or that text message is begging to be answered right away, it's important to take a long-term approach to tech–life balance. We need to be models for our children and show that actual face-to-face human interaction is still incredibly important and meaningful. More important, we need to demonstrate that safety comes first.

This doesn't mean never using tech in front of our children. Keeping our kids entirely away from technology is not the right approach either.

Barely a day goes by without there being another sensationalist blog or newspaper column decrying the dangers of technology. This kind of knee-jerk reaction toward technology goes beyond the usual sensationalist media voices. Speaking to the *New York Times* in an October 22, 2011, article, one Google executive expressed his support for banning computers from the classroom. "I fundamentally reject the notion you need technology aids in grammar school," he said. "The idea that an app on an iPad can better teach my kids to read or do arithmetic, that's ridiculous."

But this viewpoint is based in a reality that no longer exists. As technology is becoming ubiquitous, it's impossible to keep it out.

First of all, anyone who says that you should never let a young child play with a phone or a tablet has obviously never taken a baby or toddler on an airplane. When you have a screaming child on your lap and you're about to go on a long plane ride, you start to care a lot less about balance.

I also don't know what I would do without Skype. When I'm on the road, traveling for work, I live for those few minutes each evening when I can see my husband and son and connect with them face-to-face. It always seems so incredible to me that our kids are growing up in a world where they will think it's perfectly normal to talk to another person through a computer screen. To me, it still feels a little bit like magic. But our kids are going to grow up completely used to it.

My friend has a two-year-old, and once when they came over to our house, the child saw my computer, pointed at it, and said, "Grandpa!" The boy is so used to Skyping with his grandpa, he thinks his grandfather lives *inside the computer.*

Who says technology doesn't bring your family closer?

But even if it *were* possible to deny our kids access to technology, why should we? When technology is used the right way by children, it makes a positive difference in their lives. This doesn't mean technology should replace ordinary face-to-face interactions or any number of the critical learning experiences of growing up. Technology provides a powerful supporting role, enhancing those educational and developmental experiences by fostering creativity and intellectual curiosity. New innovations, along with our digitally obsessed culture, will continue to fundamentally alter education and social development. We're already seeing exciting new companies emerge in the "ed tech" (educational technology) field, and we'll see many more in the coming years.

Ten years from now, education may look completely different from how it looks today. And that's a good thing. There is no going back. So, we need to embrace these changes and collectively rethink outdated approaches to education and child development.

The reality is that preventing our children from being able to explore all the great digital resources will set them back as they grow up. No parent wants his or her child to get left behind. We wouldn't wait until our children were three or four years old to speak to them or let them hear language. So, if technology is the language of the future, why would we wait to introduce our kids to the tools that are going to define virtually every aspect of their lives, their relationships, and their careers?

While the iPad is not a babysitter, it's also not an enemy. It's amazing to me that at such a young age my son can already navigate the phone so well, play with apps, and swipe across the screen. If he didn't know how to do that, and all his classmates did, wouldn't I be doing him a disservice? The sooner children get their hands on technology and start familiarizing themselves with it, in a responsible and supervised way, the greater their chances of understanding, engaging, and thriving in the world.

After a recent panel discussion I participated in on this topic, I was moved by an e-mail I received from a friend in Phoenix. She included a photograph of her five-year-old foster grandson and asked, "Is this child already behind?" She talked about the parents she sees who aren't introducing their children to technology early on, because they have less disposable income to spend on tech, because their children spend the majority of the day in day-care centers that don't have tech, or because the parents simply don't know or care enough to prioritize having devices in their homes.

She makes an interesting point. While of course my conversations with more privileged families center around the question "Are my kids getting too much technology?" it is more rare, but just as important, for all parents to ask themselves "Are my kids getting *enough* technology?"

Although they do things children with more structured lives don't have time to do, like play outside with other kids on the block, they don't do the things that will help them up the ladder higher up in the middle class their parents struggled to enter. In order for these kids to avoid the tribulations their parents had, they are going to have to develop the same [tech] habits [that more privileged kids] have. The iPad and its Android relatives aren't just things that take kids away from human relationships. In the case of my foster grandchildren, they are problem-solving tutors, teaching them skills they will need to succeed in a world where their parents still struggle to compete.

—Francine Hardaway

Of course, that doesn't mean that tech should replace normal kid activities. I agree with people who say, "They're kids! They should get out and play! Just send them out into the backyard with a baseball mitt and a pile of rocks!" For young children, tech–life balance should skew *way* in favor of the life part.

It's our job as parents to not only be role models, but also help our children find the ideal balance that works for our specific families and situations. As we know from our own lives, tech–life balance is a vital skill for children to learn early on.

We live in the real world. There can be great benefits to teaching our children about different technologies and engaging with them in constructive dialogue about the pros and cons of gadgets or websites. We just need to be as mindful and conscious as possible about our own use of technology. Good digital habits for our children begin with us.

Share Wisely

The first thing we need to pay attention to as parents is how we share online.

When I worked at Facebook, I'd always correct people who assumed, years after the site had launched, that it was still something just for college kids. When we looked at the data, guess who the power users of the site were. Yes, moms. New moms spend an average of two hours *per day* on Facebook. Compare that to the forty minutes per day the average user spends on the site, and you have one wild bunch of moms who love them some sharing.

That's a whole lot of sonograms, baby photos, stroller tips, and graphic potty-training descriptions. And we've all seen it. We all have those friends who feel that no detail about their kids' lives is too small, or too gross, to share online.

I've definitely had times of being an oversharer myself. When I was five months pregnant with my son, I turned to my husband and said, "I'll never be one of those moms who goes from having her own life to suddenly only posting a million baby photos online. I have my career. I have my own life. I'm too busy for that."

"Of course you'll be one of those moms," he said to me sweetly, "and it'll be great."

Sure enough, four months later my Facebook profile was an incessant stream of baby photos, thank-you messages to people who sent adorable Onesies, and the nonsensical ramblings of a person getting only two hours of sleep at a time.

Yet there were also times when sharing proved incredibly helpful. One evening I woke up in the middle of the night feeling extremely ill. I had been having some problems with nursing and knew that mastitis was a possibility. Since it was three in the morning, there was nobody for me to call, and I was sending myself into a freak-out spiral by looking at WebMD. So, where did I turn? To Facebook, of course. And there, I was stunned by the reaction. Dozens of comments poured in offering advice and sympathy. Some tips were extremely helpful, and most important, I felt like I wasn't alone. My friends were there to help me, support me, and commiserate.

Don't get me wrong. The chronic oversharers could help us all by toning it down a notch (or ten). But now when I see something from a fellow Facebook mom that crosses the line for me, I try to put myself in that person's overworked shoes. If oversharing leads to useful advice, a good laugh, or just a sense of connectedness during a difficult moment . . . well, who am I to judge?

We've all been in awkward situations. None of us is perfect. Life is chaotic and messy and exciting, all at once. There are so many people who try to paint themselves as "the perfect parent" on Facebook that I'm actually pretty grateful for the honest ones who tell it like it is, even if it isn't always the most appetizing lunchtime reading.

I've actually seen situations where oversharing saved lives. In 2011, Deborah Copaken Kogan, a mom in New York, posted photos of her sick son, Leo, on Facebook. Leo had been diagnosed by doctors with strep throat, and after three days of medication he wasn't getting any better. One of her Facebook friends saw the post and recognized her son's swollen face as a sign of a rare and potentially fatal autoimmune disorder called Kawasaki disease, then urged her to take him to the ER. Thanks to that timely intervention from a Facebook friend, Leo was diagnosed in time to get treated and make a full recovery.

But before you post away, there are a few caveats to bear in mind. We need to remember that our children model their behavior after us. We can't expect to share every single detail of our lives with hundreds of people online and then turn around and lecture our kids when they do exactly the same thing. That makes for a poor tech–life balance and can lead to the posting of potentially damaging content. That's something we want our kids to be savvy about.

You should also remember that not everyone wants to hear these details. Chronic oversharing might lead to your friends choosing to "dial you down" in their Facebook News Feeds, so they see fewer of your posts. Maybe that doesn't matter to you, but it's something to consider.

Most important, while parents should feel free to post away, they need to remember that overshares can do more than just weird out their friends; they can shape their children's digital identities for years to come. If you post naked or unflattering photos of your child online that someone else could download or take a screenshot of, you are creating content that could potentially show up in a Google search of your child's name for the rest of his or her life.

There's a blog site called STFU, Parents, which catalogs some of the worst examples of parental oversharing on the Internet. This is just one Facebook post a parent made about her kid: "Another first tonight! Lulamonster pooped in the bath! Haha!! I'm just happy that she got all that out before going to bed."

That's not just a gross story. A few years from now Lulamonster is going to be really mad at Mom for sharing that and making it part of her online identity without her consent.

Identity begins before our children have a choice. In fact, digital identity begins before birth. From the very beginning, we need to be careful how and what we share on behalf of our kids.

Parents-to-be put a lot of effort into how they are going to announce their pregnancies online. This means our children are going to be able to look back and see not only photos of themselves as babies, but also photos from the moment the rest of the world knew there was going to be a baby.

When I first found out I was pregnant, I couldn't wait to announce it on Facebook. I spent days crafting the post in my head and imagining all the likes and comments rolling in. As the time to post got closer, I felt stressed. I realized that some of my friends might be offended if they first found out via Facebook. I made Brent sit down with me for hours to work out which of our friends should hear the news in person, over the phone, in an e-mail, or on Facebook. We had an entire matrix-like spreadsheet, which made me glad that my husband had once been an Excel whiz at McKinsey.

Once the personal calls were made, I felt immense pressure to be witty, creative, badass, and sentimental all in the same post. I didn't have a cute dog who could hold up a little sign saying "I'm going to be a big brother" or the time to paint "Due in May" across my belly, so I went with two photos.

The first was me standing in front of one of the "expectant mother" parking spaces at Facebook HQ, trying to look all tough. The second photo was me, my husband, my brother, and my brother's now-wife, Priscilla, all celebrating at a local dive bar. Mark had also just been named *Time* magazine's Person of the Year for 2010, so I captioned the photo, "Celebrating with the man of the year! (and also the girlfriend of the year, husband of the year . . . and fetus of the year! Cheers!!!)"

I wonder what Asher will think about all that. One day those are probably going to be the very first photos on his Facebook Timeline.

As well as planning a strategy for announcing their pregnancies, it's perfectly normal these days to hear expectant parents say things like "I wanted to name my child XYZ, but the domain wasn't available, so we chose a different name" or "I need to make sure I can claim a decent Gmail address and Twitter handle for my child before we tell anyone the name we've decided on." Apparently, baby naming has turned into a cutthroat land grab, where expectant mommies and daddies lay claim to valuable digital real estate before deciding on the perfectly unique, perfectly Googleable baby name.

I used to think this was weird, but I've changed my mind and now think that it is an acceptable and, actually, responsible thing to do. Our names have always been the major things someone else decides about our personal identities on our behalf. The Internet now provides the opportunity for many more people to "meet" us. So, it's more important than ever to own our names, both online and offline, as much as possible.

So, don't be afraid to share as a parent. But think first about what you want to share and how. The choices you make will echo through the Internet for many years to come and long after that baby is out of the cradle.

Becoming an Internet Jedi

All of the above is invaluable, but we still need to address the elephant in the room. How exactly should we be teaching our kids to use technology, and how do we keep them safe online? How do we achieve a balance between giving them the freedom to explore the Internet and feeling the urge to shield them from all potential dangers?

For the answer, let's go back to a time before the "Internets" as we know it.

A long time ago in a galaxy far, far away . . . (by which I mean the 1990s in Dobbs Ferry), when I was fifteen, I directed my siblings in an ambitious amateur film project. I was in charge of making an amazing space opera with a cast of loveable, heroic characters, each with its own distinctive personality and unique costume. My siblings had all spent at least twenty minutes learning their lines and were deeply committed to the project, assuming that there was enough time between dinner and homework to complete it. And none of the cast or crew was worried about a lawsuit from George

Lucas, despite the name of our interstellar production: the Star Wars silogy. It was a silogy because it was our silly remake of the Star Wars trilogy.

It was 1997 and the Star Wars trilogy had just been rereleased with updated special effects. For many kids, this was the first time watching *Star Wars*. For the older folks, this was a chance to relive happy childhood memories. So, basically everyone in the world was going crazy for Han, Luke, Leia, Chewie, R2-D2, and C-3PO all over again. It was no different at the Zuckerberg house, and once we had seen the new movies, we became even more insufferable. Now all we wanted to do was talk about light sabers, wookies, and Yoda.

Then we took our new geekdom to a whole new level. After my dad allowed us to use his old video camera, we decided to film the silogy. And I was in charge. My first director credit.

The casting was easy. Mark played Luke. Donna was Princess Leia, and once we figured out that her long hair could also double as a beard, she was also Obi-Wan. I was a little too old to be in costume, but I still insisted on getting to play Darth Vader and Han Solo. It might seem unfair for the director to award herself the two best roles in the movie, but to be honest, everyone was just glad to avoid Arielle's part. She played R2-D2.

To make it work, we stuffed her into a mini garbage can. She wasn't happy. But we all have to make sacrifices for art.

What was the best and the worst part of our masterpiece? To make the classic opening text of a *Star Wars* movie, we printed the intro message from the real movie on the ancient dot-matrix printer in my dad's office and then slowly walked the roll of paper past the camera. We didn't even try to hide the fact it was printed on printer paper; you could see the little squares at the edge of the screen.

Here's the thing: No matter how tragic the acting or how terrible the costumes and effects, that was one of the most awesome and memorable projects we got to do as kids. We wrote the script, brainstormed the scenes, and improvised all the technical stuff ourselves from start to finish. It was a huge learning experience for all of us.

In many ways, I think the Star Wars silogy was an ideal model for how to let kids learn to use technology and do something fun and creative with it. That project was a chance to experiment and create content of our own, instead of just consuming it. It was a chance to learn and explore, and also to take some risks, within a controlled environment.

Dr. Teresa Belton, a senior researcher at the University of East Anglia's School of Education and Lifelong Learning, conducted a series of interviews with authors, artists, and scientists, in an attempt to explore what effect boredom and downtime had on their lives. Dr. Belton discovered that when kids were given the freedom to waste time on their own, they filled those moments with creative projects, developing what she refers to as an "internal stimulus" that leads to a rich life of creative expression.

Turning an iPad on for your kids doesn't mean your kids have to turn their minds off. There are apps for painting, drawing, writing, composing music, reading, and scientific exploration on that platform, and they can fire up the imagination of any bored kid. Just because the piano comes in app form doesn't make it that much less of a piano.

Our parents trusted us to do the silogy project in our own way. They knew what we were doing all along and laughed about the results afterward. We didn't share the film publicly afterward (sorry), and we didn't break the camera. Also, R2 didn't suffer any lasting trauma from being stuffed in a garbage can. I think. Sorry, Arielle.

I realize now, Dad was smart to give us the freedom to experiment with technology.

When you learn to ride a bike, at first you have constant supervision from your parents, and you have extra safeguards like training wheels. But eventually you learn to ride, and the training wheels come off. When that happens your confidence and ability increase, and your parents have greater trust in you to ride safely. You no longer need to be supervised at every moment, and you get to ride to the end of the street without permission. As you get older and more competent, you can ride downtown to buy candy at the store or go to a friend's house. Eventually your parents trust your judgment enough that they don't have to worry about where you're going at all. Okay, okay—parents will always find *something* to worry about. But at least they can breathe a bit easier knowing where you're going and what you're doing in your spare time.

That's how we need to think about technology with our children. We need to help our kids develop good habits, but we also need to give them the freedom to take controlled risks and learn to be independent. We should always be at least aware of what they're doing online, but we shouldn't have to stand over their shoulders and constantly watch them.

Striking a balance requires work, and we need to approach safety as a conversation. That conversation begins by teaching our kids about the wonders, as well as the dark sides, of the Internet. The web provides amazing knowledge, entertainment, tools, and services, and passing on an interest and proficiency in how to use these things is one of the most valuable gifts you can give your child. But our children should be able to recognize that the Internet can be a dangerous place, too.

Online, people can be mean and hide behind anonymity to bully others. There are unsavory sites and people online, and there is a lot of unreliable information. Just as our kids learn to read, they need to learn digital literacy: how to sort through the good and the bad online. And they need to learn about what is right and wrong to post and share online, about themselves and others.

They're only going to learn if we know enough about technology to teach them. After one post on my Dot Complicated newsletter, about how to talk to your kids about posting revealing pictures online, one mother wrote me a personal e-mail with a troubling story. Recently, her seven-year-old daughter had come home from school saying, "Mom, can I show you what a girl in my class did today?" At which point she showed her mom that the girl in her class had uploaded naked photos of herself to Instagram. Not only that, she had uploaded naked photos of her mother getting out of the shower. Did I mention this girl was seven years old?!

This particular mother knew what Instagram was. She was able to sit down with her daughter and talk to her about the photos, asking her, "Do you think what your friend did was right?" and having a meaningful conversation. After which, she promptly called the other girl's mother to tell her what was going on. The other mom had no idea what Instagram was or how a naked photo could even get onto a cell phone in the first place.

How can we expect our children to engage safely with technology if we don't even know enough ourselves to talk to them about it? I know we're all busy. We all have exhausting professional lives and home lives. We have homes to maintain, friendships to maintain, and some of the lucky ones still have bodies left to maintain. Alas, there's now one other area to maintain: our knowledge of the latest technology, gadgets, apps, and sites our children are using.

There are some very important rules to set and conversations to have with your children about online behavior.

1. Your body is your business only. Think before you post revealing pictures.

2. Don't bully or go along with other people who are bullying.

3. Only add "friends" online if you also know them in real life.

4. Always treat others with respect, the way you would want to be treated.

5. If you're going to put something in writing, make sure you would be comfortable if it was reprinted in a newspaper.

6. Only say something to someone online if you would also say it to that person's face in real life.

7. Be careful with personal information about yourself or your family. Only share things with people you trust.

8. Be vigilant against predators, lurkers, and bullies.

9. Above all, guard yourself and your dignity, and stay safe.

Of course, just as the training wheels come off the bike when our kids are ready, we need to give kids the freedom and confidence to explore the Internet for themselves. We can't just program them to do the right thing; they have to learn what the right things are as they grow as individuals. The conversation needs to be two-way. We need to have an ongoing conversation with our kids about safety by staying as engaged and involved in their lives online as we would offline. But our kids also need to know we trust them to explore the web themselves, to make mistakes and learn from them—and that if anything ever goes wrong, we'll always be there for them.

So, when your kids are just starting to build an online identity, begin by letting them go online only on a main "family" computer, kept in the living room or another shared family space. This will help you keep an eye on everything at the beginning. This is also why I don't want Asher running off with my iPad. I'm happy for him to play on it, but if he gets into trouble, I want to be there to help him. (Also, as a toddler, he'll probably smash it.)

As kids get older, let them build up to the privilege of going online unsupervised, whether on a laptop in their rooms, on their mobile phones, or anything Wi-Fi enabled, such as an e-reader or an iPod. Collect computers and gadgets or turn off the Wi-Fi in select areas of the house after a certain time at night. Install monitoring software for your kids' laptops, tablets, and phones. Because tech evolves so quickly, any apps I mention here will likely be outdated by the time this book is published, so I'd encourage you to find websites that explore these topics. Or you can always join our newsletter community if you're interested in staying up to date with the latest innovations.

Most important, though, build up trust with your children and educate them. By the time they're accessing the open Internet, make sure they're ready for it. There's no app out there that can teach the type of good judgment parents can teach.

It's worth noting that many social media sites require that children be thirteen years old to have an account, due to the Children's Online Privacy Protection Act (COPPA). That doesn't mean that some kids don't find ways around it or lie about their age to create an account before their thirteenth birthday. It does mean that the U.S. government and these sites recommend thirteen as the age when it's appropriate to have social media accounts. Certainly that's something to take into consideration.

Another way to keep your kids safe is to limit the time they spend on websites that permit anonymous interaction. We need our kids to learn early the benefits of authentic identity online.

While bullying is part of human nature and has always existed, the issue is exacerbated on the Internet because people hide behind anonymity. It's easy to be mean when you're using the name "bumblebee57" instead of your real name and nobody knows who you really are or can hold you accountable. It's also easy to be mean simply because it's so easy to type something and hit "enter." You don't need to invest a lot of thought in the effect your words have on others, especially when they're strangers. People forget that when they post something mean, there is a human being on the other end reading the post. It's so easy in this anonymous, digitally passive-aggressive world to forget that.

Recently, I was having dinner with a friend who runs a popular blog with an A-list actress. During this particular dinner, she told me about how the actress recently said, "I don't think I'm going to go on Twitter today, because I don't feel like seeing people tell me to die."

That's why I've always been an advocate for people using their real names and identities online. This was a decision we made early on at Facebook. We quickly saw that when people used their real names, they were much less likely to write horrible things, because their identities were attached to everything they said! They couldn't hide from their nasty words, so they were much more thoughtful before writing or posting.

The lesson for our families is that, as part of the conversations about safety we have with our children, we need to talk about the importance of authentic identity in staying safe online. When we participate in the networks and services where we can interact as ourselves, we are generally operating in the clear light of day and can stay away from the dark side of the web.

Of course, there are always exceptions to every rule. When you are yourself online, you can invite abuse or harassment that begins in the offline world and follows you online, or vice versa. Because we have one integrated identity these days, there can be no respite from a determined bully. And, in a sense, sharing in the digital age means we're always on a real-time performance review, every second of our lives. I used to turn to only my closest friends if I wanted to hear brutally honest opinions about myself. On the Internet, I get that from everyone. So much so that whenever I go home, the *last* thing I want is more harsh criticism. From my family, I seek out more nurturing, support, and positivity.

There are ways we can deal with this. One is to teach our kids to respect other people online just as they would want to be treated themselves. If we do our part, we can make the Internet a better place for everyone. As we know, good behavior can be contagious.

Daniel Cui was a freshman high school goalie on the varsity soccer team living in Hillsborough, California. After he failed to stop an opposing team's game-winning shot, a small group of students began posting photo albums of him on Facebook with the caption "Worst Goalie Ever." Daniel was embarrassed and humiliated and didn't want to go to school. But then Daniel's friends and supporters rallied. Another group of students began changing their profile pictures to ones of Daniel as a way of showing solidarity with him. Hundreds of students went on to express their support.

The next day, when Daniel walked into school, he wasn't afraid of the bullies anymore. Being himself online meant his friends were on his side when things got rough.

Tips for Achieving Tech—Life Balance in Your Family

Work. Sleep. Family. Friends. Fitness. Pick Three.

Sometimes the demands of work must take priority over the needs of our children. Sometimes the demands of family mean there's no time for our friends. Sometimes the demands of our friends mean there is no time for the gym. Don't beat yourself up about it. Focus on the few things that are important to you, each day. The beautiful thing about life is that each morning is a fresh, new start, and your priorities can change from day to day. We can't be great at every single thing every single day. As long as it pretty much balances out in the end, it'll all be okay.

Good Digital Habits Begin at Home

Kids mimic the way you behave. If you want your kids to have good digital habits, you must set the example. If elbows were once banned from tables for being impolite, now perhaps it should be cell phones. If your kids see you texting and driving, chances are they'll do it too. The technology will always be a part of their lives, and it's your job to teach them how to live those lives.

As your kids are experimenting with technology, don't be afraid to experiment with them. Sign up for the new sites they're using. Experiment alongside them. By staying knowledgeable about current technology, you can keep up with what your kids are doing—and help keep them out of trouble.

The iPad Is Not the Babysitter

Technology can entertain, educate, bore, and amaze our children. But it will never be a substitute for the benefits of dedicated human interaction. It's a tool to be utilized for good and then put away when it's no longer needed. And if you're not raising your kids, someone else is. Don't trust the Internet to do it for you.

Digital Identity Begins Before Birth

Digital identity no longer starts at birth. Even before our kids are *born,* we can post their ultrasound photos online. Make sure you proactively secure your child's digital identity as early as possible. Register e-mail addresses and a .com domain for your kid, and at least Google your baby's name once before choosing it. A more common, conventional name will be less Googleable than an exotic one. But don't get carried away by this process. Kids need love, not search-engine optimization.

Stay Safe Online—Be Authentic

Online safety is not something that happens with a single setting of certain filters. It is a conversation that takes place over a long period of time. Your children should feel empowered to come to you when there's a problem, not feel intimidated that they're always being watched. Setting up a family computer in a common room or maintaining a "white list" of appropriate sites could be ways to keep your family safe online. Limiting kids' time on sites that permit anonymous interaction may also help cut down on their exposure to bullies.

In the end, the Internet is a web of people, not computers. And it's up to us to keep it safe.

Another set of tools can play an invaluable role in keeping our kids safe online: parental controls and privacy settings. Most sites have privacy controls, and it's our responsibility as parents to familiarize ourselves with them, so we can help our children use them properly.

Don't be afraid to set up parental controls on your Internet browser or put limits on the sites your kids are looking at. It is, after all, still your computer! Feel free to define an acceptable list of websites your kids can visit, or at the very least a list of websites they need to avoid.

If you want your kids to get the maximum benefits out of controlled risks, they need the risks as well as the controls. Your default attitude toward Internet browsing should be toward openness.

At the same time, you need to stay on top of the latest fads and phenomena, so you can make informed choices about what's safe for your kids. As parents, you have a responsibility to know what's going on, and it pays to keep an open dialogue with your kids, to read the tech blogs occasionally, to check your kids' Internet history, and to share knowledge with other parents. Don't let your kids keep you in the dark. And when new websites or apps crop up that "all the cool kids are using," have a discussion with your child about how to use them safely and wisely, before you give the okay.

Giving your child his or her own cell phone or computer is no longer a luxury; it's a necessity. Many people ask me when it's appropriate for a child to have a cell phone, and I usually respond with somewhere between the second and sixth grades. It's different for every family, but once you have play dates, after-school activities, and extended periods of separation, these devices become tools for ensuring safety, essential for both you and your child.

It's also incredibly important that we teach our kids how to effectively configure their privacy controls online and to make use of reporting functions, so if they get into trouble online they can get help from administrators or other trusted authority figures. You want to make sure that your kids know how to share content with the right groups of people and to keep strangers out, just as they would in their offline lives. Of course, understanding the value of authentic identity, and spending time mostly on services that are based on this concept, will eliminate some of these problems.

We also need to give our children, especially teenagers, some credit and not project our own tech insecurities onto them. In many focus groups we conducted at Facebook, we found that teenagers were actually far more savvy about utilizing privacy settings than one would think. A study by the TRUSTe privacy management company recently showed that about 70 percent of teens know how to effectively configure their privacy settings on social networking sites, which is more than the number of parents who know how to do it.

It's funny. When kids are young, we constantly tell them to share. Share your toys, share your games, share your snacks. But the second we turn to the digital world, our instinct is to protect and close off. *Don't share!* we cry. Don't share photos, don't share information, don't share passwords. The very nature of sharing has become complicated and muddled.

At the end of the day, parenting your kids online is really just parenting, period. There's no one formula or correct solution to all our problems. But learning about the risks, rewards, and challenges technology creates in the lives of our children and families will ensure that they are equipped to come of age in the digital world.

The Internet can be a highly valuable place for kids. It doesn't have to be scary, and we shouldn't try to keep our kids offline. Start a conversation with your kids about safety, and make sure it's an ongoing dialogue. You can always find the dark side of anything if you look for it. Let's remember to stay focused on the light.

 CAREER

A Life More Authentic

It was a freezing cold Friday night in Davos, Switzerland. As I struggled to take off my UGG boots and walk through yet another metal detector, I grumbled and, for a moment, wished I was back in California with Brent. *Keep going,* I thought. I was far from home, but this was not an experience to pass up.

In January 2010 I was invited to represent Facebook at the World Economic Forum in Davos, which is an annual event that gathers heads of state, CEOs, celebrities, academics, and the media to discuss innovative solutions to some of the world's most pressing problems.

The forum hosts a Friday night Shabbat dinner, which is the traditional meal that begins the Jewish Sabbath. I'm very proud of my heritage and was honored to be invited this year. Plus, I could finally tell my mom I found a place to go for Shabbat.

The dinner took place in a small room at a hotel just outside the conference center. As I walked in and looked around, I was quickly overwhelmed. All around me were famous and important figures from the Jewish world. Elie Wiesel, the Holocaust survivor and humanitarian was there, along with Shimon Peres, president of Israel and Nobel Peace Prize laureate, Gary and Laura Lauder of Estée Lauder, the head rabbi of Moscow, the under-secretary of state to Hillary Clinton, and so on. The number of accomplished and influential people in the room that night was mind numbing.

I was a little nervous but soon put at ease. Despite the impressive company sitting around the tables, it was really just a regular Shabbat dinner—intimate, humble, and welcoming.

As I was settling into my seat next to Julius Genachowski, chairman of the FCC and a leading voice in the promotion of net neutrality, and getting to know my other equally impressive tablemates, I felt a tap on my shoulder. I looked up and saw Yossi Vardi, one of the world's most beloved Israelis and a connector of all things tech, Israel, and Jewish.

"Randi," Yossi said. "Elie Wiesel usually does the opening song for the Shabbat dinner, but he isn't feeling well tonight and would prefer not to."

Go on . . . I thought.

Yossi continued, "Will you sing for us, 'Yerushalayim shel Zahav'?"

I paused for a moment to take in what Yossi had just said. "Yerushalayim shel Zahav," which means "Jerusalem of Gold," is a popular Israeli folk song and a tribute to the beauty of the country. I thought back to the times in high school when I had sobbed my way through *Night,* Elie Wiesel's book about his survival in Auschwitz. I had never thought I would even meet someone like Elie Wiesel, let alone sing for him. And to stand in front of the president of Israel, as a young Jewish woman, and sing about Israel was a tremendous honor.

"Yes," I said confidently. "I'd love to." And a moment later, confidence gave way to panic, when I realized I didn't really know the lyrics. Minor detail.

I called Brent. He answered the phone, groggily awakening from a deep sleep. It was the middle of the night in California. "Brent!" I said. "I need you to text me the lyrics to 'Yerushalayim shel Zahav.' No time to explain."

Within a few moments, the lyrics were on my BlackBerry. I quickly leaned over to the FCC chairman sitting next to me, handed him my flip camera, and said, "Here. Please record this, or my mom will totally kill me."

I soon found myself standing next to President Peres, who introduced me to the room.

"In San Francisco," he began, "there is a young Jewish boy by the name of Mark Zuckerberg. He runs a big company called Facebook that today I think is worth $100 million, or something very modest. Tonight his sister is here, but she's not Facebooking. She's singing, and she's going to sing for us."

I babbled about how excited I was to be there, looked down at the lyrics on my BlackBerry, smiled at how lucky I was to have such an amazing husband, took a deep breath, and began to sing.

"Avir harim tsalul k'yayin, vereiyach oranim,
Nissah beru'ach ha'arbayim, im kol pa'amonim."

(The mountain air is clear as wine, and the scent of pines is
carried on the evening wind with the sound of bells.)

"U'vtardemat ilan va'even, shvuyah bachalomah
Ha'ir asher badad yoshevet, u'velibah—chomah."

(And the sleeping tree and stone is captured in her dream. The city
that sits alone, and at its heart—a wall.)

The room was silent. I took a breath and asked everyone to join in on the chorus. In a beat, the room filled with voices, all singing:

*"Yerushalayim shel zahav, v'shel nechoshet v'shel or
Ha lo lechol shirayich, Ani kinor."*

*(Jerusalem of gold, bronze and light, behold, for all your songs, I
am a harp.)*

As I finished singing, I paused to look around the room, holding back beams of pride, for myself and for Israel and the Jewish people. I could practically hear my mother kvelling all the way back in Dobbs Ferry.

I walked back to my seat, half glowing, half trembling with nerves. After a moment, I excused myself from the table, called Brent, and broke down in tears. Words fail to describe how I felt, standing before some of the people who had made the State of Israel possible and, three generations later, singing about Israel to them.

Later that night, I posted the video on Facebook (all credit goes to Chairman Genachowski—he tapes a mean video!) along with a note about how, as a young Jewish woman, this was one of the most meaningful things that had ever happened to me. It was an important moment in my life, and I wanted to share it.

Within minutes, comments and messages were pouring in, publicly and privately. I got messages from people telling me how moved they were by the moment. Some people shared stories of their own past Shabbats. Others asked personal questions about my Jewish heritage and reflected on the difficulties of staying in touch with their own. People I had never met before commented on how inspiring it was that I was willing to stand up proudly before a room and celebrate my culture, my heritage.

I was so grateful for all the messages I received. I felt connected on a scale that I've rarely felt, and not just with friends and family but an entire community. I felt warm and loved.

But others had a very different reaction. Privately, I received a number of messages from friends and mentors I respected telling me that I shouldn't have done that at all. It wasn't professional, particularly for a woman executive, to put myself out there like that and to sing in public. And to sing on touchy subjects like Israel or religion was just too risky.

I didn't feel the need to defend myself at the time, but later I spent a lot of time wondering if I had screwed up. On one hand, I'm proud of my Jewish heritage. It is an essential part of who I am. But I was at Davos to represent Facebook. If my singing reflected poorly on the company or me, then maybe it was a mistake.

Also, it felt as if it wasn't just the fact that I sang that had raised some eyebrows. People wondered why I had to record it and post it online. Maybe it would have been okay to sing if I had just kept it to myself?

I was conflicted. That moment was one of the most important of my life, and my Facebook profile is a reflection of who I am. How could I not post the video? Something was all wrong in the land of the Internet.

In 2011, I went back to Davos, this time pregnant with Asher. For months leading up to the conference, I hoped I would be asked to sing again, but I was wracked with nerves about what I would say when asked. When they did ask, as desperately as I wanted to say yes, I decided to stay focused on my career and I declined, citing my pregnancy as an excuse. They were very gracious, and were even worried about me, but I felt hollow and sad, as if I had compromised my authentic self for my professional identity.

I resolved then to be true to myself both online and offline, in private and in public. And that might demand a shift in attitudes and preconceptions about what a "professional" is, but this is something we all have to do. Because that's when you get to really belt your heart out.

The 360-Degree Professional

This philosophy was put to the test when I had Asher. Even though I promised myself many times over that when my son was born I wouldn't become "that mom" on Facebook, I fell hard off the wagon and committed the typical new-mom-on-Facebook crime of documenting and posting every waking second of my new baby's life. First yawn? Adorbs. Facebook it. First hiccups? Obviously all my friends want to see that. First spit-up? Share. Snoozing in a crib? Snoozing in a stroller? Snoozing in a park? OMG, soooo cute! Who wouldn't want to see baby photos fifty times a day?

Well, I soon found out.

I had some pretty honest coworkers, and one day one of them decided to give it to me straight and called me on the phone to say, "Randi. Asher is adorable, but you've gone off the deep end. You can't keep posting a zillion baby photos. You have a professional reputation to uphold as well. Do you really need to post about every single motherhood moment on Facebook?" We laughed about it, and the conversation was soon over. But when I hung up the phone, I felt horrible again.

By posting those baby photos, I was being true to where I was in my life. My days were entirely consumed by Asher, at the time, and it didn't seem unreasonable that my Facebook Timeline would reflect that. I was being completely authentic online and much more than I had been in a while.

What if uploading photos of my son was compromising my professional "brand"? Would it really damage how colleagues and business partners on my Facebook friends list viewed me? Or was my online profile supposed to be a sterile, "professional" version of myself, in which I only showed brief glimpses of my life outside of work? And if the solution to these challenges was to share less, did that mean I should stop being authentic or try to be less authentic? If less, how much less? Twenty-five percent less? Fifty percent?

Ultimately, after a few days of thinking about it, I decided it was okay to blend my personal and professional life online. I am now convinced that the people who think we need to create a purely professional, one-dimensional brand online have got it totally wrong.

Hear me out.

Today, the people who work with me or do business with me get to read my newsletters about the future of the entertainment and technology industries, see the latest behind-the-scenes photos from the Zuckerberg Media studios, and watch the various interviews and media appearances I do every week. But they also know that I love my son and that I love to post lots of photos of him. They know that I love to travel. And they know that I sing in a cover band called Feedbomb (the illustrious successor to Evanescence Essence).

If before I could be professional Randi at work, motherly Randi at home, and social Randi around my friends, now I have to be all-of-the-above Randi. In the era of smartphones, social media, and authentic identity online, it's no longer possible to separate your personal and professional identities. It doesn't matter if I use LinkedIn for my business self and Facebook for my social self and CafeMom for my mommy self. If someone wants to track down any of these identities, all they need are some basic Googling skills and everything could be revealed.

Right now, there are two generations in the workforce who think in diametrically opposite ways about identity. Executives who came of age in the pre-smartphone era take it as a given that you should have a separate professional persona that reads like a profile in *Forbes* and doesn't overlap with your personal persona. But my generation came of age in a world with social networks, and we know that we don't have that luxury anymore. We understand that the business leaders of the future will be three-dimensional personalities, whose lives, interests, hobbies, and passions outside of work are documented and on display.

Instead of trying to maintain one narrow, unrepresentative version of our professional selves for our colleagues, the best leaders will have 360-degree identities, in which the personal and the professional will be combined seamlessly.

Do my colleagues, professional contacts, or online acquaintances think any less of me because they've seen more than just professional Randi? I hope not. We're multidimensional, multifaceted people, all facing the same challenge of reconciling our personal and professional identities. We're more than just our jobs. We're more than workers or bosses. We're mothers, fathers, sisters, brothers, friends, spouses, singers, poets, politicians, foodies, and sports fans, all at the same time. We always have been. That's what makes us so awesome.

So, we should embrace it. The answer isn't fewer baby pictures; it's more baby pictures. It's not that I should post less; it's that everyone else should post more.

If someone doesn't want to do business with me because I post photos of my son or my band, well, then maybe they're not the type of person I want to do business with either.

Let's change what it means to be professional in the Internet age, and let's admit what's already obvious: the time when your personal identity was a secret and you needed to keep it separate from your work is pretty much over and done.

This is a good thing. By choosing to be myself, I can eliminate worries my colleagues may feel about opening up. And there's no reason that any of this makes me a less effective manager or partner. If anything, being my authentic self online makes me a better leader at work.

Research has shown that when you refuse to share personal details on Facebook with your colleagues, it reduces your likeability in the office, when compared to people who do share. A white paper released by researchers at the University of Pennsylvania's Wharton School showed that because information exchange leads people to form stronger bonds with one another, people who shared personal information with their work colleagues and bosses, and seamlessly blended their offline and online lives, were thought of as better workers. People like to work with someone they can relate to.

Of course, there is a place for LinkedIn and other strictly professional social media sites. I am personally a huge fan of LinkedIn and find it to be an incredibly powerful platform. But until it can accurately capture the full picture of who I am, the 360-degree Randi, it will always be just a piece of my online identity rather than the soul and lifeblood of it.

So, going forward, what do we do? How do we make the most of this new terrain and avoid being embarrassed or overwhelmed?

Living an authentic identity online that includes our work lives is possible with a little common sense, a few risks, and a deft mastery of privacy controls. If social media is making mini celebrities of us all, we might as well be our own publicists.

First off, since we're all going to be exposing more about ourselves online in our careers, we need to start being a bit more tolerant of what we learn about our colleagues and professional contacts. Employees are people too, and most (or at least *some*) of their lives are spent outside of the office. Pretending that people don't have lives and interests outside of work is a ridiculous approach to take. As the distinction between public and private behavior changes, so should our expectations of one another.

Second, just as our personal lives increasingly blend into our professional lives, our professional lives will also blend into the personal. What happens when e-mails from the boss start flooding in at eleven P.M.?

According to a survey by Right Management, nearly two-thirds of workers say their bosses e-mail them over the weekend and expect a response. And in a survey of U.S. workers by Good Technology, more than 80 percent of people said they continue working after physically leaving the office. On average, people claimed to work an additional seven extra hours each week from home. That's basically another full day of work. That same survey revealed that 68 percent of people check their work e-mail before eight in the morning, 57 percent check it on family outings, 38 percent routinely check at the dinner table, 40 percent do work e-mail after ten at night, and 50 percent find it difficult to put their phones down to go to bed, admitting to reading and responding to work e-mails long after climbing under the covers.

In this new always-connected scenario, we need to be sure that we respect one another's tech–life balance, just as we would our own. I always make a point to tell my new hires, "You will get e-mails from me really late at night. I do not expect you to respond to them late at night. That's just when I work best, so that's when I attack my in-box."

If your boss does expect you to respond to e-mails at any hour of the day or night, any day of the week, and it's negatively affecting your life, you may need to have a conversation that sets some rules and boundaries, based on a sense of mutual respect of everyone's tech–life balance.

Given that your work colleagues will also be your friends on social media, there could come a time when you are still waiting on a response from them relating to a work matter and see that they have had time to post something on Instagram or make a move in Words with Friends. At these times, it's important to keep in mind that everyone has a life outside of work that they're entitled to. Just because someone is doing something online, it doesn't mean they're always working. If we want to achieve tech–life balance ourselves, we have to grant it to our colleagues and friends alike. If you don't get a response to your e-mails or texts right away, cut the other person some slack.

This also applies in the world of our friends. With texting, in particular, it *feels* as if we should get a response instantaneously. But demanding an instant reply to your messages is like tapping someone on the shoulder and interrupting a conversation they're having with someone else. Unless we want our friends to be always alone, we shouldn't maintain an impossibly high standard for replies.

Give people time to get back to you, and don't feel pressure to get back to someone immediately if you're paying attention to someone else. We need a kind of social detente in the arms race to reply faster and faster. The smartphone makes it theoretically possible to be connected to everyone instantaneously, but the human mind can't handle that kind of demand.

There's a software company called FullContact in Denver, Colorado, which has started giving their employees a $7,500 bonus if they don't take their phones with them when they go on vacation. The reason? When you briefly disconnect from being always online and restore some of the tech–life balance in your life, you return to work refreshed, invigorated, and more productive.

Studies have borne this out. According to a 2006 internal study of their employees, Ernst & Young found that the year-end performance ratings of their employees improved by 8 percent for every additional ten hours of vacation the employee took.

If all else fails, you can always move to Brazil. In November 2012, Brazil's President Dilma Rousseff approved legislation that said checking your smartphone for work e-mail after hours qualifies you for overtime pay.

When you maintain a single authentic identity online, as well as a proper tech–life balance in the workplace, and grant the same to your coworkers, you not only become more well-regarded and trustworthy at work, you also become more productive.

Plus, who doesn't like baby photos of kids on vacation? There's nothing more adorable than watching little Asher try to walk in the sand. I think I have a picture of it, somewhere.

Building an Online Brand

Posting personal photos or videos online is only half the story. The Internet lets you do so much more. You can advance your career and career prospects by harnessing the concept of an authentic online identity and establish for yourself an online "brand."

You can't just wake up one day and declare yourself an expert, though you might not realize that from the thousands of people who refer to themselves as things like "influencers," "thought leaders," "social media legends," and "idea accelerants." It takes time, energy, and results to build credibility. There are lots of talented people out there who truly *are* experts, but most of the time you'll find those people don't need to declare themselves as such.

Building a personal brand doesn't have to be a negative thing or turn other people off. You can harness the tremendous broadcasting power of social media to tweet, post, and blog about new ideas and developments related to your field of interest and make yourself known as a smart, interesting, and ambitious person.

Access is all-encompassing. If your boss can see your baby photos, he or she can also see your blog posts. So, why not make those posts count for something?

This is a capability that we've never had before in our careers. The sophisticated search tools and social networks, the massive increase in Internet users over the past decade, and the ease of building and engaging with communities of real people online make it easy to find an audience. If you can't blog, tweet. If you can't tweet, then retweet. Build a name for yourself. Some of the Internet's most successful blogs were built up from an audience of zero. Everything from the Daily Kos to the Drudge Report started out as a one-person operation. If even one of your posts goes viral, you could secure a permanent readership.

And even if you work hard on a blog and still can't attract traffic, if there's a link to that blog on your Facebook page, your employer is going to read it, and that may be all the audience that matters.

Perhaps at one time, photocopying and handing your bosses and fellow employees a self-produced newsletter about the goings-on in your industry would have been seen as a little too ambitious, and maybe a little weird. Now leaving those same items on your blog for your workmates to discover will communicate the same message without the awkwardness. You don't need to schedule a thirty-minute one-on-one with your boss to brag about your achievements. You can just go and achieve.

I'm always impressed by initiative and ambition. I do worry, however, about how my company could be negatively affected by something an employee of mine does online. This is why I believe it's critical for employers to empower their employees to use social media freely, but to also train them how to use it wisely.

Most companies wouldn't dream of having an executive go on TV or do a press interview without hours and hours of media training and prepared talking points. Well then, in a world of smartphones, where every single person in your company is speaking to a public audience, why wouldn't you train your employees as well? Why would you just send them out into the world as potential mouthpieces for your company without arming them with some skills and a few key things to say?

Empower your employees to be good ambassadors for your firm. Instead of just having one corporate identity, remember that your employees are part of that too, and they can help strengthen and augment it online, plus humanize and develop it into a living, breathing brand.

Social media skills are going to become necessary in the new job marketplace. Employers are going to want to hire people who know how to use social media, rather than those who ignore it or are bad at it or do not appreciate its power. Every employee who is online is now a kind of PR representative for his or her firm. A smart employer will use this talent to its advantage, rather than just see it as a liability and try to silence it.

The same holds true for whole companies. Back when Facebook was getting started, companies would usually hire a college student to run their Facebook pages and manage their social media presence, if they bothered to have one. It was an afterthought to their real marketing efforts.

Nowadays, senior marketing teams have dedicated, full-time professionals managing companies' offline social media. There's even a new kind of job that didn't exist a few years ago: online community manager, which is an Internet-savvy marketing and customer-service position. These professionals are responsible for helping the Old Spice guy respond to people on YouTube, contacting customers who have posted bad Yelp reviews, and calming angry Twitter storms, to find out what went wrong and make everything all better.

Now that everyone has a megaphone, people have started shouting. The only way for companies to respond is to go out and listen—and to know how to start a conversation.

What to Do Now That Your Voice Matters

I learned a great lesson about the newly democratized power of words online when I first began to step out onto the national stage.

It was the fall of 2008, a pivotal moment for American politics and social media. Facebook had just passed the point of being thought of as a thing college kids used to Poke one another and was emerging as the vital platform for political discussion, analysis, and candidate-liking that it is today. As Facebook's marketing lead, I was in charge of coordinating our response to the Republican and Democratic national conventions. We were careful to treat both conventions equally, as well as both candidates, John McCain and Barack Obama. As a place for people to share their lives, Facebook had to be a neutral, nonpartisan platform. My job was to make sure both parties could easily call on Facebook support for their online campaigns and convention coverage. So, in the last week of August, I duly made the trek with my team to Denver, for the Democrats, and then a week later to St. Paul, Minnesota, for the Republicans.

Although we resolved from the beginning to treat both events equally, my experiences in Denver and St. Paul were night and day.

At the Democratic National Convention, social media folks were in high demand. Everyone wanted to meet with us, we were asked to do a lot of press, and we were invited to every hot party. "How should we be thinking about social media?" "How can we use Facebook to connect to voters and donors?" In every room I walked into, I was peppered with questions by hungry young activists and community organizers. And I stood less than ten feet from Kanye West at his private performance at the Google–Vanity Fair party. The atmosphere was hopeful, optimistic, curious.

In contrast, the Republican convention was somber from the start. The party was meeting just days after Hurricane Gustav had devastated the Gulf Coast states, forcing the main conference proceedings to be pushed back by a day. A lot of the party's grassroots members were deeply ambivalent about McCain too. But there was also just a general apathy toward Facebook, social media, and technology. People didn't want to hear about what we had to offer and weren't interested in making use of the tools and resources at their disposal. I was only able to secure a few meetings with middle-ranking party officials. I ended up spending the majority of my time e-mailing from my hotel room in a futile effort to get people to meet with us, before finally giving up and taking my colleagues Adam and Simon to the Mall of America.

A few weeks later, after the election was over, I sat on a panel organized by a venture-capital friend of mine named Dave McClure, discussing the interplay between technology and politics. When it was time for the audience Q&A, I was asked whether I saw any big differences between the way the RNC and the DNC thought about Facebook and reacted to social media.

I paused for a moment, thinking back to the two vastly different experiences.

"At the Democratic National Convention we were like rock stars. Everyone wanted to meet with us," I said. "At the Republican National Convention I sat in my hotel room by myself for three days. No one would meet with us. I was begging people to meet with us."

A blogger recorded that statement and uploaded it to the Internet. A few days later—in the comments section on YouTube—Matt Burns, the RNC's communications director responded. "With all due respect, Randi Zuckerberg is totally full of sh*t on this one," he said, before accusing me of liberal bias.

None of this might have been a big deal had it not been picked up by blogs, one of which ran the headline "Randi Zuckerberg is Totally Full of Shit." I won't list the comments that appeared in these stories, but suffice it to say, they focused on my weight, appearance, and all sorts of alleged personal and character defects. After I saw this outburst of aggression and viciousness, I was so upset that I cried for a while in front of my computer. I wondered how I could show my face at work the next day, or whether I would even have a job to go back to.

In the end, the ruckus received only a passing mention at work. The storm passed, and I didn't get fired. Most people were actually very supportive, and I learned some important lessons from that moment.

For starters, the Internet may be the world's new watercooler, but it's also the world's most efficient, perpetual outrage machine. It's entirely possible that something you do or say, even if trivial or completely innocuous, can set that machine off.

The *New York Times* motto is "All the news that's fit to print." If the Internet had a motto, it might be "F*** it. Write whatever you want." Anything that might generate clicks is fair game to be posted, whether or not it's truthful or accurate. Personal insults traded between people online are quality "click bait," moments of outrage built up by writers and bloggers purely as a way of generating more article views. This is a model of "journalism" that has evolved over the last few years as part of the monetization strategy of all the usual tabloid suspects.

For example, earlier this year when I announced that I was writing a book, one blog posted a short article discussing the news. The article was snarky and clearly designed to feed the trolls, who responded with gusto. Here are just a few of the comments that followed the article:

"I have nothing to hear from this woman."

"Who the s*** cares about Randi (Jesus, what a name) Zuckerberg?"

"If my Facebook-addicted sister reads this book, I will murder her and her entire family."

Nice. But even if none of this abuse remotely fazes me anymore or affects the work I'm doing, the outrage machine has plenty of ability to destroy careers and reputations.

Google the words "Internet" and "fired," and you'll come up with an endless list of people who have had their employment-related gaffes put through the ringer of the Internet-forwarding machine. There are untold numbers of people with video résumés, employment-seeking letters, stories of misbehavior at major corporations and law firms—anything and everything has, at some point, made it to the rounds of the forwarded.

In 2012, Gene Morphis was the chief financial officer of Francesca's Holdings Corporation. He maintained a personal blog and Twitter account, and like any good Internet citizen, liked to post about his life, including things of interest to his company, such as "Dinner w/Board tonite. Used to be fun. Now one must be on guard every second" and "Board meeting. Good numbers = Happy Board." His company began to take an interest in his online life. The tweeting CFO continued, "Earnings released. Conference call completed. How do you like me now, Mr. Shorty?"

As it happens, Mr. Shorty apparently didn't like him tweeting information that could affect the stock price of his publicly traded company. Morphis was fired.

Why do people, even in senior positions and with plenty of experience, make such dumb mistakes with social media? The Internet seduces us to share our most intimate, private, and crazy thoughts, because it's so incredibly easy and because it sometimes makes us forget we're communicating and interacting with actual people. When all you're doing is talking to a computer, that requires a lot less work than looking someone in the eye and saying the same thing to that person's face—and a lot less nerve.

The answer to this, as I've discussed throughout this book, is all about being true to who we really are, both online and off.

Certainly in the pre-Internet world, there was a lot of face-to-face human cruelty that didn't require a computer screen to rob people of their shared sense of humanity. There always were and always will be some bullies on the playground, mean bosses at work, vocal critics, and just plain jerks. But when so many of our personal interactions play out via faceless interchanges with a global reach, the temptation to hurt and the potential to be hurt are exponential.

Keep this in mind when posting online reviews of businesses. The rise of social media means that companies are far more accessible to us, as consumers, than they have been previously. There was a time when, if a server at a restaurant treated you poorly, the best you could do was complain to the manager. If you were extremely influential, perhaps there was a slim shot your opinion about a restaurant would appear in a newspaper or as a snappy "three-word summary" in a Zagat guide. Now, a single tweet about a single bad moment can set off a tweeting avalanche that won't abate for days and days. If a single server doesn't meet your expectations, you can go straight to the Internet and influence the perceptions of thousands through Yelp.

At any given moment, my Twitter feed usually contains at least one person complaining about a travel delay or bad customer service. (Damn you, Delta, how could you do this to me? Don't you know who I AM?! I have a really high Klout score!) One of my all-time favorite viral YouTube videos is "United Breaks Guitars," in which a passenger sings about United Airlines breaking his guitar during one of its flights. As I mentioned earlier in the book, both positive and negative feedback travels faster and farther than ever before. Which means that the average Joe can now have the same impact on a business as a professional reviewer can.

This is a tremendous amount of power for consumers to have. For the first time, customers and clients have the ability to render their verdicts on *us* in the public sphere, and our reputations depend on impressing a far wider audience than just the boss. As discussed earlier, with the posts our friends put on our walls, our authentic identities aren't merely what we curate; they also consist in large part of what others say about us. The moral here is, to the best of your ability, be nice to the people you meet since, in effect, there is a kind of performance review that takes place in real time every day online.

And keep in mind that just because you have the megaphone doesn't mean you always have to be shouting into it. If someone does something to offend you, be careful of going straight to the Internet. Your response should be in proportion to whatever went wrong in the first place.

I've spent a lot of time recently thinking about the trend of social media "shaming" and how it's gotten wildly out of hand. In March 2013, at a Python programming language developer's conference in Santa Clara, California, a woman overheard some men behind her making off-color jokes about dongles, which are a kind of laptop cord. They also joked about "forking" someone's code. What happened next illustrates how quickly things can escalate online. Instead of confronting the dongle jokers directly, she made an example of them by snapping a photo of them and calling them out on Twitter for being inappropriate. A few thousand retweets later, and the dongle jokers were fired. Then, news of that firing hit the Internet, kicked up the outrage machine, and she found herself fired too.

That's the world we live in. One tweet, one photo, one blog post is now all it takes for people to lose their jobs, their reputations, and their credibility.

Another case is that of Lindsey Stone, a girl who, as a joke, took a photo of herself holding up a middle finger inside Arlington National Cemetery, and she posted it on her personal Facebook page. Bad taste? Yes. Worth losing her job over? Unclear. However, an angry group launched a "Fire Lindsey Stone" Facebook page, started a Change.org petition, and harassed her employer publicly until her employer let her go.

So, what's the takeaway? If you see people doing something you don't like, and you talk to them privately about it, you're giving them time to think about what they did and change their behavior the next time. But when you immediately take their behavior public, you've summoned the Internet hordes, and there's no telling when or where they'll stop. Empathy—that essential aspect of face-to-face interaction—seems to be especially absent when it comes to online mobs.

There's no doubt that it's uncomfortable and awkward to confront someone. It's much easier to just point, click, and share than it is to actually deal with speaking to another human directly. But wouldn't you want a person to talk to you before posting, if the situation were reversed?

When you shame-post someone, it can have serious effects on that person's life. A friend of mine recently found out just how far her voice would carry online when she rage-tweeted on the UPS Twitter account that she didn't get the package she was expecting, even though she was home, because the delivery guy didn't ring her doorbell. Not long after this tweet, the same delivery guy showed up at her door again, deeply apologetic for his failure to ring the doorbell. As she later posted on Facebook, "I didn't know it would trickle down to the poor guy that delivers all our stuff. *Ugh.* I need a little perspective sometimes—feeling sorry."

Most of us have been guilty of this in the past. I admit that I've overreacted and called people out for things on Twitter instead of telling them how I felt directly, and I sometimes pretend to be taking pictures of something when what I'm really doing is trying to get a picture of something else funny right behind it that I can post online. But recently I've been thinking and reflecting on my own behavior. It hurts to be mocked, whether online or offline, and if you wouldn't want it done to you, don't do it to others. Just because there's only a slim chance the person will ever find out, that doesn't mean you're not being a bully.

Maybe in a world where everyone is a critic we all just need to toughen up a bit and grow thicker skin. But I feel that's not quite the answer either. What we need are better unspoken social rules and etiquette around this sort of thing. Just because we all have the functional equivalent of a printing press at our fingertips, and in our pockets, that doesn't mean we need to use these devices to settle scores.

With great power, according to the eminent philosopher Spider-Man (and Voltaire, of course), comes great responsibility. We cannot expect the Internet to solve all our personal problems, nor, given the exponential power of our voices online, should we turn to the Internet unless necessary. When you choose to publicly shame someone on social media, it's as if you're telling that person that you don't think they're worthy of a second chance. Most people deserve a second chance. So, we need to choose when to use the megaphone, when to address our problems directly with the people with whom we have issues, and when to just let it all go.

Tips for Achieving Tech—Life Balance in Your Career

Keep Things in Perspective

Sometimes social media can ignite a firestorm, which has devastating, career-ending consequences. But most of the time, getting things wrong online merely results in temporary embarrassment and some hurt feelings. In the end, most things pass, and you have to keep them in perspective. Winston Churchill said that if you're going through hell, keep going. If you're going through a flame war, stop, drop, and roll with the punches. In time, people will move on and you'll recover.

It's Okay to Friend the Boss

A Wharton School study, titled "OMG, My Boss Just Friended Me," showed that people were anxious to friend their bosses because of the potential to suddenly eliminate what was previously thought of as a solid hierarchy of communication. If it used to be hard for you to talk directly to your boss, this often means that it was as uncommon for your boss to talk directly to you. Social networking can level this distinction with a simple friend request.

What matters, then, is using the platform to your advantage. Add the boss, but make sure to utilize your privacy settings and only give access to things you *want* him or her to see.

Protect Your Privacy

Become a privacy-settings expert. It might seem tedious, but it will pay off. Of course, the controls are never foolproof. It may also be smart to practice posting abstinence. If you stand to lose a job or friends if certain aspects of your behavior were brought to light, then it's probably smart to stop either doing these things or posting about them, and be cautious of other people posting about them, too.

Digital Posts Can Have Real-Life Consequences

A few years ago, I was going out in NYC with my girl crew when, like so many before us, we were rejected from this super-trendy, speakeasy-themed artisanal cocktail bar in Chinatown with a famously strict door policy. Waving us to the side, the bouncer told my friends and me that we didn't fit the "dress code," which clearly meant "not slutty enough." Annoyed at the bouncer, I pulled out my BlackBerry to vent my frustration on the then-new platform called Twitter.

"Worst bar ever = apothecary in NYC. Worst bouncer ever = james. It would be a huge bummer if their facebook pages 'accidentally' went down."

I thought it sounded vaguely funny *at the time,* but I didn't invest a whole lot of thought into my message. My friends and I continued on, hit up a few hot spots in the nearby Lower East Side, and called it a night. When I woke up the next day, my

Google Alerts were at red alert, and my in-box was filled with howls of outrage from bloggers and commentators complaining about my threatened retaliation against the club.

Ah, man.

I had no idea the Internet could propel my single tweet so far. Something I meant as a bad joke had taken on a life of its own. Of course, I didn't have the desire or power to delete anyone's profile, but as someone with a perceived influence at Facebook, that was an incredibly dumb and irresponsible thing to tweet.

I didn't quite get it then, but I do now. When going out for a night with friends, it may be a good idea to assign, for the night, a "designated poster," a friend with whom you have to clear any online post before hitting "share." If you've had a couple of drinks and are about to go on a posting spree, this could help prevent you from making a potentially career-ending mistake. Better yet, just enjoy the moment and don't post at all until the following morning.

The Goldilocks Problem

Knowing what to post when the workplace could be watching is a kind of Goldilocks problem: you can't be too hot or too cold. Some people may overshare *everything* and risk getting into trouble; others may overcompensate and share nothing. Both approaches are bound to fail.

The answer is to share, but know your limits. There isn't going to be a world where we can have our online cake and eat it too, where we can say anything we want online and expect it to have zero ramifications in the offline world. That wasn't true of the things we said before the Internet, and it's certainly no more true now, when our passing thoughts and bad jokes have their own universally accessible archive.

When I was at Facebook, I had a very talented young intern, who may not have known that I was a Twitter follower of hers, and I could see all her tweets about partying, drinking, and . . . let's call it "college living." Entertaining as they were, her tweets put me in a difficult position as her manager. I couldn't tell her what to post on Twitter. I didn't own the account and didn't have the right to ask her to stop tweeting. But by the same token, people knew she was working for me, and I felt that her behavior online was beginning to reflect on me as her boss.

I ended up having a sit-down chat with her, saying, "If you're going to tweet, be thoughtful. Please don't make yourself—and don't make me—look bad." She agreed, and everything was cool.

This may seem obvious, but it can't be said enough: always think before you tweet. If you're doing something that can get you fired, or is illegal, don't post about it. That doesn't make you inauthentic; it makes you smart. Sometimes the safest option really is abstinence. If you are at a job where you live in mortal fear of the HR department, keep that in mind when posting *anything* online.

If you're applying for any job with a large company—and increasingly, companies of any size—don't imagine that posting updates to do with #sex, #drugs, #moredrugs, and #evenmoredrugs won't have some kind of effect on your employment prospects. Remember that an authentic version of yourself lives online, but there may be some things you're better off not letting anyone know about.

Share what matters to you, with the friends that matter. Sharing in the workplace doesn't mean sharing everything. Your boss doesn't want to see your spring-break photos. People just want to see the generic stuff: baby photos, FarmVille updates.

Actually, scratch that. No one wants to see your FarmVille updates.

· · ·

In 2012, I was invited back to the World Economic Forum in Davos, this time no longer as a representative of Facebook but as the CEO of a media company that now bore my name. And once again I was asked to sing at the Shabbat table, this time a song called "Shalom Aleichem."

This time, I said yes without hesitation. And as I took the microphone, I looked around the room and saw many familiar faces. I thought of my childhood, growing up in Dobbs Ferry. I thought of Brent and Asher, my own family waiting for me back in California, and of the people who could not be there and whose lives we were honoring on what also happened to be Holocaust Remembrance Day.

I sang, and as the room joined in, the moment we shared was touching and all too brief.

I had come a long way. In the course of the past few years, I had learned what it meant to take risks in my personal life and my professional life. And I had learned the most important secret of all: that both those lives are part of the same thing. I knew there would be no way to separate the two, so the one life I lived might as well be authentic.

 COMMUNITY

I Get By with a Little Help from My Friends (and Kickstarter)

In this chapter, I want to talk about one of the hardest but most re-warding things on social media today: doing something good for other people and the community.

We've talked a lot in this book about how technology can make our everyday lives and relationships more complicated, whether in our friendships, our romantic entanglements, or our careers. As all those tricky situations and painful stories show, these aren't unimportant problems. If we refuse to take the perils of technology seriously today, there can be serious consequences for our lives and livelihoods.

We must keep things in perspective. Technology, as a tool, can be used for both good and bad. And in these final pages, I want to talk about some of the opportunities that exist for us to help create some good in the world.

My first experience of this was in 2007, amid tragic and terrible circumstances. On the morning of April 16, 2007, a student went on a shooting spree at Virginia Tech in Blacksburg, Virginia. Thirty-two people were killed and seventeen wounded before the gunman took his own life. It was the worst school shooting in U.S. history.

In the days that followed, something very poignant began to happen on Facebook. Many of us who shared a similar sense of anguish and sorrow began to find a common way of expressing ourselves. My Facebook News Feed began to fill with notifications of friends changing their profile pictures. They were all changing them to the same image: the letters *VT,* the Virginia Tech insignia, placed over a black ribbon. In a sudden spontaneous outburst of emotion, people from across America and many other countries were making a small gesture of solidarity with the victims and against the horrors of gun violence.

In time, the black ribbons faded and people's profile pictures returned to normal. But I was struck by what I had seen.

Facebook began as a place for trivial and everyday interactions, a place for college kids to Poke one another, post photos of their lunches, and procrastinate instead of study. What began in the aftermath of Virginia Tech showed that Facebook and social networks were transforming into something else, a place not just for personal expression but collective action—and today it's just one example of what happens when people on the Internet rally behind a cause.

Social media give us all a voice, and at certain incredible moments, we see hundreds, thousands, or millions of people bringing their voices together to form a chorus. As people take a stand and show their support for a cause online, their friends join in, followed by friends of their friends, and so on. This is word of mouth at scale. Together, all those people become a force for change in lives and communities all over the world, for so many different causes, in so many different ways.

Over the past few years, we've seen the Internet revolutionize fund-raising for charities and good causes. Every November there is the awesome silliness of "Movember," when men compete to grow the most ridiculous moustaches possible and then proudly display their whiskers online to raise awareness and funds for the fight against prostate cancer. On our friends' birthdays, we occasionally see generous souls "donating a birthday" to a charity. And there are many apps devoted to fund-raising, sometimes on a grand scale. An app on Facebook called Causes has connected over a hundred million people with three hundred and fifty thousand causes and raised over $30 million for everything from curing cancer to stopping genocide.

We've also seen citizens and consumers rally to change the behavior of powerful governments, corporations, and institutions and to make the tools of communication serve the interests of people who have been historically ignored.

In March 2012, Bettina Siegel, a Houston-based mom and blogger, started a petition on the website Change.org that asked the USDA to stop the use of "lean beef trimmings" as filler in ground beef destined for school meals. Beef trimmings are also known as "pink slime," something that's probably not good for kids. Just nine days after the petition was launched, over two hundred thousand people had signed it, and a new public debate had begun over the potential health risks of pink slime. The debate ended when the USDA announced that, starting in the fall, it would offer school districts a choice of beef either with or without the filler.

A single concerned mom started a movement that confounded an entire industry and its army of lobbyists, and she achieved what health activists and celebrity chefs had spent years calling for.

Social media's role in driving political change is well documented, as discussed in chapter 3. Over the past few years we've seen citizens mobilize on an unprecedented scale using social networks, mobile phones, and the Internet, and when people cry out for change today, as much attention is paid to the people on the streets as on the web—and often the former begins with the latter. Witnessing the Arab Spring, Iran's Green Revolution of 2009, anti-austerity protests in Europe, and the U.S. general election of 2012, we have all seen countless examples of how online action has driven incredible offline outcomes.

Of course, it's fair to ask whether most of the time changing your profile photo, liking a cause on Facebook, contributing a few dollars to a friend "donating" his or her birthday, or retweeting a nonprofit's request for money actually does anything, or if it's just a cop-out form of slacker activism, or "slacktivism." It's also fair to ask how widespread online movements really are.

Scientists and researchers have asked these questions. And the data is very clear. The impact of social media is not a myth. Word of mouth at scale has a *really big* scale and creates tangible real-world impact.

During the midterm congressional elections of 2010, UC San Diego researchers and Facebook's data science team ran a study on how social media affected voter turnout. At Facebook, we already understood the immense power of social suggestion, and as the election approached, we often talked about how a News Feed story saying which of our friends had voted would probably be far more effective in getting Facebook users to the polls than just another boring PSA explaining how it was everyone's civic duty. I helped create a banner that appeared in the News Feeds of sixty million people on Election Day with a message encouraging them to go out and vote, a clickable "I voted" button, and a list of their friends who had already clicked on the button. In the study's control group, the users saw a banner with only the go-out-and-vote message and no social information.

After the election, researchers checked publicly available voting records to see who in these groups had actually voted. The results? People who saw the "social" message were about 2 percent more likely to go out and vote than those who saw only the generic message. That might sound small, but those couple of percentage points equates to another approximately 340,000 people. In tight races, or with controversial initiatives on the ballot, those votes might make all the difference.

And in the years since, the power of the Internet to amplify people's voices has only grown. In March 2013, the U.S. Supreme Court met to begin debating the future of gay marriage. On Monday, March 25, Human Rights Campaign, a prominent LGBT advocacy group, called for people to change their Facebook profile pictures to a pink-on-red equal sign as a show of support for marriage equality during this historic debate. The group posted the image to its Facebook page and invited people to use it.

By the next day, millions of people across America, and across the world, were proudly displaying that image, or some variation of it, on their profiles. When Facebook researchers dug into the data, they found that within twenty-four hours roughly 2.7 million more people in the United States had changed their profile pictures than on a regular day, or about 120 percent more than usual.

This is what happens when people use technology today to stand up for their beliefs. Millions of people stand with them, and the effects can be awe-inspiring.

A study by the Pew Research Center in February 2012 showed that, on average, every post we make on Facebook can potentially reach an audience of over 150,000 people through friends of friends. When you post something online, your voice can echo through the Internet and spark new conversations. All those conversations have the potential to lead to new ideas, and ideas to action. We are the most empowered generation in history. And if you have the power to change the world, why wouldn't you?

Start Small

So, how exactly can we create change that has a massive impact? How can individuals use technology to create movements that reach thousands or millions of people? I will try to answer that by going back to 2010.

On January 12, 2010, a massive earthquake struck the Caribbean nation of Haiti, just sixteen miles from the capital, Port-au-Prince. In less than a minute, thousands of homes and buildings collapsed, hundreds of thousands of people lost their lives, and millions of people were made homeless.

As news of the disaster began emerging late on January 12, I was a world away—metaphorically and literally. I was standing in the lobby of the palatial Venetian hotel in Las Vegas, attending a Stanford business school event. My husband was at Stanford's Graduate School of Business at the time, and every year they have a tradition of all the business school students dressing in '70s costumes and heading to Vegas for one night. There we were, surrounded by people wearing polyester jumpsuits, disco jackets, and comically large lapels, as images from Haiti began to stream in on CNN.

I knew that I had to get home as soon as possible and mobilize my team at Facebook to help with communications. I ducked into a hotel lobby café with Brent and, with two phones and a laptop going simultaneously, we frantically tried to rebook my flights and contact my team.

It was in the middle of that café that I got a call from the White

House. (I still get excited saying that. I mean, who does that happen to besides Jack Bauer?)

It was Macon Phillips, the White House director of new media: "Randi, I'm reaching out to all my contacts in Silicon Valley. We need all the major tech companies to help with Haiti. We need Facebook to do something." The White House was launching a massive relief and recovery effort for Haiti and was trying to mobilize donations and resources from across the United States. They were working with many different partners already, who were planning big online fund-raising drives, but they had a specific vision for using Facebook. "You guys have the most people online, and it's the only way we're going to get to a critical mass of people fast. Can you help us with raising awareness about all the different campaigns that are going on?"

Whoa.

I headed to the airport fast. I needed to get back to the office right away. Thankfully, there was in-flight Wi-Fi, so all the way back to San Francisco I was conferencing with my colleague Adam Conner. And by the time I landed, the two of us had sketched out a plan for a Facebook page that could serve as a hub for getting out information about the relief and recovery efforts.

Over the following hours, a lot of people at Facebook worked with me to turn that idea into a reality, working through the night fueled by adrenaline and an urge to do something good. The next morning, we launched the "Global Disaster Relief" page, a clearinghouse of information for individuals, nonprofits, governments, and everyone else involved in the Haiti response, to find out the latest about the relief efforts and how to support them. It was less than twenty-four hours since the earthquake had struck.

By this time, people on Facebook were posting more than fifteen hundred status updates about Haiti every minute, and organizations such as the Red Cross and Oxfam were raising hundreds of thousands of dollars in donations through their own Facebook pages. Using the "Global Disaster Relief" page, we started working to amplify these efforts to a broader audience. Later, we also tried other tactics, organizing awareness-raising online town halls with everyone from the United Nations to Linkin Park and fund-raising with the sale of virtual gifts in the Facebook gift shop. Already, within a day, we were part of a much broader constellation of groups online working to try to help people.

When I think about what makes an effective online social movement, I see an important lesson in that moment. All change *starts small.* The online campaigns for Haiti ended up mobilizing millions of people and dollars in donations and having a huge impact on the lives of those suffering from the tragedy. But none of these efforts started that way. They began with individual phone conversations, e-mails, and messages asking for help, which quickly snowballed into something much larger. No one at Facebook who worked on the "Global Disaster Relief" page had any idea how popular the page would be or how many fans we were going to get. We didn't care. We knew this was something we wanted to do, so we focused on just launching the page as quickly as possible so that we could start being useful. And in the end, we had an impact.

Sadly, there are too many charities and causes that struggle with this. During my time at Facebook I often talked with groups that cared only about getting millions of fans or likes. They invested more time in driving up these numbers then actually thinking about what the real purpose of their campaigns were or what kind of value they wanted to provide people with their pages and content.

What's a better approach? More important than focusing on driving up your number of fans is making sure you have the right people.

The White House was able to mobilize the Internet community over Haiti by calling people at key Silicon Valley firms. I was only able to get the Facebook page going by having a team of extraordinary people who were willing to work through the night to launch it.

There are plenty of other examples of giant acts of change that started off with just small groups of people really committed to a cause. In January 2011, Wael Ghonim was a marketing executive for Google based in Dubai. When the Arab Spring began, Wael returned to Egypt, his home country, to take part in demonstrations for greater political freedom. Wael became the creator of a popular Facebook page that mobilized many Egyptians to take to the streets in protest, and today he's credited as one of the most important and courageous young leaders of the Egyptian revolution.

But Wael almost didn't get to play his part in the revolution. A few days after street protests, he was picked up by the police and quietly thrown into a jail cell. None of his friends or family was told of his arrest. Wael could have expected to sit out the rest of the revolution from his cell.

Thankfully, however, just before being arrested Wael had managed to update his Twitter account. "Pray for #Egypt," he tweeted. "Very worried as it seems that government is planning a war crime tomorrow against people. We are all ready to die #Jan25." Because of that worrying message, concerned friends and family began to scour local hospitals and prisons looking for Wael, and an online campaign formed to raise awareness about his disappearance. The posts and stories continued for twelve days until Wael was released from prison.

Wael went on to play a much larger role in the revolution. But it wasn't the power of the online masses or the value of the Internet as a platform for new ideas that got him released. It was something much less lofty: the fact that Wael had used social media to communicate with his friends, family, and closest supporters—the people who would never give up on him and who went the extra mile to find and free him.

When a massive tornado hit Oklahoma in May 2013, social media enabled organizations and responders to mobilize quickly and effectively. Hashtags on Twitter such as #OKNEEDS provided valuable information on available places to stay, get a bite to eat, or even charge a cell phone with a dead battery. Reddit, a site that played a large role in mobilizing Internet users to discover the identity of the Boston Marathon bomber a few weeks earlier, now played an important role in displaying images of missing persons and lost items displaced during the tornado, so they could be returned to their rightful owners. Social media enabled people around the country to feel like they could do something rather than just sit by helplessly.

According to a 2012 study by the Red Cross, 76 percent of those who found themselves in the midst of a natural disaster during the previous year used social media to contact friends and family; 44 percent turned to social media instead of calling 911 and asked their friends to contact help lines or responders on their behalf; and 37 percent used social media for help finding shelter, supplies, and support. In the aftermath of Hurricane Sandy, twenty-three Red Cross staffers monitored over 2.5 million Sandy-related tweets and social media postings.

So, sometimes you don't need a million people to make a difference. You just need a few people who really care. That's why all online movements for social good need to focus, first of all, on building meaningful connections and long-lasting relationships with a strong core group of supporters.

Just as important is making sure that when you ask people to lend you their support online you don't just ask them to like your page or passively consume information. To turn a small group of people into a movement with a much larger impact takes hard work and concrete action. As much as possible, you need people to participate—to actually do something.

It can be as simple as inviting people to share a meme or change their profile pictures, attend events, or contribute ideas and content through YouTube videos, tweets, or blog posts. But there are other more ambitious ways of getting people to make a tangible action.

One example is how Barack Obama totally reinvented the model of electoral fund-raising during the 2008 campaign. Instead of a traditional campaign strategy that focuses on getting large corporate donations, the Obama 2008 campaign ran a massive online campaign encouraging people to offer small donations and get their friends to contribute in kind. Over 90 percent of the contributions made to the Obama campaign were for less than a hundred dollars, but the cumulative impact of all these small sums completely turned the race on its head. It also probably laid the groundwork for a lot of crowdfunding sites that followed soon after, such as Kickstarter, Crowdrise, and Indiegogo.

Kickstarter is a fantastic and powerful example of how you can get people to create change themselves. Kickstarter is a website that allows people to raise money for creative projects through crowdfunding, collecting many small investments from people in order to create original films, theater, music, games, and inventions. Before Kickstarter, you could be a "patron of the arts" only by donating tremendous sums of money to your local museum or at an opera house gala. But now anyone can become a patron of the arts. And they have. Since launching in 2009, nearly 4 million people have pledged over $588 million to fund more than 40,000 projects, from the Pebble e-paper watch to 10 percent of the films at the 2012 Sundance Film Festival. And in 2013, the documentary *Inocente* became the first Kickstarter-backed production to win an Oscar. So, by enlisting the support of millions of Internet users, artists and entrepreneurs are helping create new innovations and forms of cultural expression.

One final point on starting small: sometimes when organizations and causes are starting off, they guard their independence jealously and see their work as a struggle to define themselves against other "competitors" working to drive action on the same issues or for the same community. I'd encourage them to think otherwise.

To get to scale necessarily means working with others, and if you're working to drive action on an important cause, then there will inevitably be other organizations working to achieve the same things you want. If you really care about the change you want to achieve, swallow your ego and work out how to identify the right partners among other organizations, and then work with them to achieve change together.

During my time in Silicon Valley, I worked on several projects that were truly pan-Valley, which is unique for such fiercely competitive tech companies. During the election, I worked closely with Google and Twitter. On Election Day 2008, we actually displayed on Facebook users' home pages a Google Map that showed the location of their nearest polling stations. Postelection, I worked with execs from Google and Twitter to teach a seminar at Stanford's business school. And Haiti, of course, was an unprecedented moment of cooperation for the industry. I sat at a FEMA roundtable convened at Facebook with senior executives from companies around the Valley. And I worked closely with Google to get out the message about their Person Finder app.

So, if all those industry competitors could come together to do good, there's no reason others can't. And for groups that are just starting out, sometimes the best thing they can do is work with other organizations in the same field to share resources, cross-promote, and help drive audiences and traffic to one another.

Of course, no matter how desperate the crisis, eventually public attention will begin to fade. There's a natural limit on the human capacity to remain engaged, especially when people have their own challenges to cope with. The trick is to keep a passionate community and user base mobilized and interested in a cause after it has faded from the public consciousness and long after everyone has stopped donating to the Red Cross and gone back to their normal lives.

This is where effective storytelling comes in.

The Power of Storytelling

In 2010, Facebook reached the incredible milestone of five hundred million active users. The company wanted to mark the occasion in some way, and internally there was a lot of discussion about how best to do it. A big party? A video about Facebook's evolution as a product?

It was an important moment for the company and a testament to the work of a lot of amazing people. But I didn't think that the focus of the celebrations should be on the company. I always held the belief that the most powerful thing about Facebook wasn't just the platform itself, but what people were doing with the platform. The average users didn't care all that much about how many people were on Facebook or how big our server farms were. They cared about the value the site added to their friendships and relationships.

At lunch one day in the cafeteria with my colleague Matt Hicks, we fantasized about driving a bus around America and talking to real people about how technology was changing their lives for the better. From that conversation, Facebook Stories was born. In the end, we never got a bus, but we did decide to build a Facebook app that would shine a light on the individuals, communities, and causes that were being empowered by the site and help them tell their stories to the world.

Drawing on my experiences, I knew that in addition to showcasing the lives of fascinating people, it was also important that the app showcase best practices for other brands looking to market themselves on Facebook. I had to set a good example and create something that could be replicated by others on the platform. So, Stories received no internal support or special engineering resources. The whole app was programmed using third-party developers anyone could work with and promoted with ads anyone could buy.

We worked with the design firm Jess3 to create an app that was able to highlight user stories and plot them on a map of the world. The app was also embeddable, so, for example, the Red Cross could put it on their Facebook page and ask for stories about disaster relief, or Babies "R" Us could use it to collect and display stories about social media and childbirth.

One of the first stories shared on the app was that of Ben Taylor, a seventeen-year-old boy from Kentucky who used Facebook to rally support to rebuild his state's oldest outdoor theater after it was damaged by flooding. There was the story of Holly Rose, a mom in Phoenix, who was prompted by a friend's Facebook post to examine her breasts for suspicious lumps, found one, and then sought treatment for cancer. She used Facebook for support during her cancer treatment and went on to organize an awareness-raising campaign on Facebook called "Don't Be a Chump! Check for a Lump!"

There were countless stories of lives changed, saved, and empowered thanks to connectivity.

Facebook Stories even came to the attention of Justin Bieber, who posted a link to the app on his website. Subsequently, this unleashed a torrent of tween "Beliebers," who posted so many "stories" about how they loved Justin and wanted to marry him that we had to temporarily shut down the app. Though many of my colleagues stopped being Beliebers that day, I was impressed.

All in all, the project was a huge success. The stories told by our users were inspiring, poignant, and memorable. And looking back, Facebook Stories was one of the most important and human additions to the Facebook platform that I worked on.

At the core of a person's life is a narrative. This is why, if you're trying to convey a message, simple, relatable storytelling will enable your audience to see themselves through the eyes of the people you are trying to help. This is the beginning of empathy. The story is the glue that binds together the human experience; it's the device through which both understanding and a desire to help emerge.

An unintended side effect of the many methods now used to broadcast for charities and social causes is tragedy fatigue. Like many of you, I am constantly getting requests for Kickstarter and Indiegogo drives, Kiva microfinance pitches, Causes.com causes, and any number of other microsourced and crowdfunded campaigns. During the 2012 election, the pitches from interest groups and the Obama and Romney campaigns were even more unbearable. There was clearly no limit on the number of e-mails that campaigns would resort to sending, and all of these were impersonal in the extreme—they knew my first name but nothing else about me.

There are only so many times people can ask you to support a crowdfunded Crowdrise campaign or help them "donate" a birthday to charity or donate to a race, walk, marathon, half marathon, racewalk, or SantaCon, before you start turning down every request.

Effective storytelling breaks through the crowd to reach a receptive audience. With a good story, it's possible to grab not only people's attention but also their understanding.

If an organization is trying to raise money for hunger, it's not enough to just say that a certain number of kids go to bed without food every night. It is critical to relate on an emotional level with the struggle and plight of someone who may be in a foreign country and with whom the audience might not feel any natural connection.

The one charity e-mail request that, for me, broke through the pack came from a friend of mine who was raising money for a multiple sclerosis charity walk. Normally, these kinds of things all blend into the background or stay relegated to the limbo of the never-replied-to Facebook "event." But this one request included a link to a video, which I clicked on. It showed the guy at his wedding, lifting his mother, who suffered from MS, from her wheelchair in order to dance with her while uplifting music played in the background. It was one of the most heart-wrenchingly emotional moments I've ever seen on the Internet. After I wiped the tears from my eyes, I immediately clicked to donate.

There's a reason that one TV ad, which pans slowly over sad, abandoned pets while a Sarah McLachlan song plays in the background, makes me scramble for the remote, or that Proctor & Gamble "Dear Mom" ad that played during the Olympics made me bawl my eyes out. They tell effective, empathic stories and do so using mostly music and imagery.

This isn't to say that the best videos will necessarily be the weepiest ones. What matters is the story. Effective pitches weave a narrative that centers on the struggles borne by an individual or group and touches on themes of our common humanity and our desire to overcome life's challenges. Online fund-raising, at its heart, requires telling the story of an individual, one person, and connects with the person on the other side of the screen.

It's for this reason that I remain impressed with the *Kony 2012* phenomenon. This was a brief, yet powerful moment where social media were used to rapidly escalate global awareness of the plight of kidnapped child soldiers in the Lord's Resistance Army, a rebel militia fighting in northern Uganda and surrounding countries under the direction of a man named Joseph Kony. In March 2012, a group called Invisible Children, Inc., released a thirty-minute film about Joseph Kony and promoted the film across social media using the hashtags #makekonyfamous, #kony2012, and #stopkony. The cause soon went viral. Within a short period of time, the movie had tens of millions of views and the attention of the world.

It wasn't long, however, before *Kony 2012* faded from view. Many organizations that had been trying to raise awareness of Kony for years felt slighted by the attention paid to these newcomers. No viable plan was ever organized to stop Kony. Then, one of the filmmakers had a public breakdown, and the organization never really recovered. Joseph Kony remains at large, and "Kony 2012" is now a kind of cynical shorthand for a social-media-based awareness campaign that looks grandiose but accomplishes nothing.

However, *Kony 2012* actually accomplished quite a lot and deserves to be praised. The Invisible Children group took a humanitarian crisis that hardly anyone knew about and, within a matter of days, leveraged the power of Internet culture and social media to inform the world. They showed that it could be done. Perhaps they have yet to accomplish the ultimate goal of stopping the Lord's Resistance Army, but not even the United Nations has managed to do that yet.

The key takeaway is this: the reason *Kony 2012* became so well known so fast was because the YouTube movie told an effective story. The movie didn't just say there were children suffering; it showed you. It humanized what would have otherwise been just another foreign news story. Have people made films about the suffering occurring in far-off places before? Yes, but no one had ever done so leveraging the power of the Internet to inform so many, so quickly. Perhaps the group was not equipped to handle the scale of the movement they had started, but that is no reason to discount what they did accomplish.

As for what the next Kony 2012 will be, all bets are off. Anything is possible.

Think Global

In the age of the Internet, distance is no longer a barrier to assistance. It's just as easy to donate to a cause based thousands of miles away as it is to donate to someone standing right in front of you. Just go online, click "donate," and you're done. Or send a text. That's it.

What this means is there is no such thing anymore as a purely local cause. What in the past may have been the concern of only a small, local community—a drive to fix the church roof, for example—can now be the concern of the entire globe. When we promote causes that resonate with universal human sympathies, dreams, and desires, we can move the entire world.

Take, for example, the story of Karen Klein. Karen was a bus monitor for middle school children in upstate New York. In June

2012, she was the recipient of some harsh verbal abuse from the kids on her bus. Using their smartphones, they filmed themselves calling her, among other insults, a "f***ing fatass" and uploaded the video to the Internet the next day, under the title "Making the Bus Monitor Cry." Soon, the video racked up hundreds of thousands, then millions of views. As the video went viral and attention increased, the kids responsible for the abuse were punished by the school district and made to apologize to Karen.

The Internet community didn't stop there. Within days, an Indiegogo account was set up on her behalf by a complete stranger to her, with the aim of gathering $5,000 to send her on a vacation. Within a month, the crowdfunding site raised over $700,000. With some of that money, Karen set up the Karen Klein Anti-Bullying Foundation, and she retired on the rest. What once would have been the concern of a few local school board officials was now the concern of the entire world.

So, there's a lesson here for everyone using the Internet to drive awareness for a cause. When the world can hear, be prepared for the world to listen—and be ready to enlist its support. With every cause, it is vital to think globally. This means turning a local story into a global story, and a local issue into a global issue—something people will take an interest in even if it's outside the realm of their everyday lives and communities.

In late 2010, I joined the Global Entrepreneurs Council of the United Nations Foundation. The council brings together young leaders from many different industries, civil society, and the media to help find new solutions to tackle some of the biggest problems facing the world, from war to poverty to climate change.

I joined the council to drive action on initially one issue: malaria. Malaria isn't a major national challenge for the United States, and it is not something that the average American necessarily thinks about. But by focusing on telling stories about just how devastating malaria is for communities in sub-Saharan Africa and Southeast Asia, our campaign aimed to tell a universal story—about people and societies struggling to overcome great challenges—and to tap into the natural goodness and kindness of Americans when faced with people in need. Through conferences and town halls broadcast online, we explained to audiences the death toll caused by malarial infection from easily preventable mosquito bites, and we provided moving first-person testimony from malaria survivors. As part of this, I also organized a malaria prevention town hall at Facebook, the first for a charitable cause. During this, I shared a personal story about a young man I had known from Stanford's business school, who had fallen victim to malaria and died during his spring-break travels. This sad story showed that malaria isn't just a tragedy for distant lands; it's something that affects many people from communities all over the world, from all walks of life.

In the end, anything that speaks to universal struggles of hope or loss—struggles to survive, to get an education, or to live a good life, free from fear, abuse, and suffering—has the potential to find a global audience.

In January 2012, the social media profile photos of millions of people all across the world "went dark." They turned completely black in protest of the threat of online censorship posed by two proposed pieces of American legislation, the Stop Online Piracy Act (SOPA) and the PROTECT IP Act (Preventing Real Online Threats to Economic Creativity and Theft of Intellectual Property Act, or PIPA). Other global organizations, such as Global Voices online and Wikipedia, also disabled their home pages in protest. Even though SOPA and PIPA were strictly American pieces of legislation, because they affected the Internet, they potentially affected the entire globe, and so they got the world's attention. The cofounder of Wikipedia, Jimmy Wales, explained that they had disabled their home page "to send a broad global message that the Internet as a whole will not tolerate censorship."

SOPA and PIPA quickly failed. By the next day, most of the bill's cosponsors had deserted the legislation, and many more lawmakers from both Republican and Democratic parties had come out to voice their opposition. All this was because of the giant outpouring of online opposition on that day of action, swelled by millions of people and organizations from around the world.

If you ever wanted to rally the world around a cause, now is the time. Just give people something to do. Of course, in order for a cause to resonate with the world, it needs to be understood by the people of the world. So, we must speak a common language.

Sometimes a message can be spread with more than just words. Take, for example, the story of "Dancing" Matt Harding, a twenty-nine-year-old software developer from Connecticut who shot to Internet stardom in 2005 after he made a funny and moving video of himself dancing in different countries around the world. He went on to make two more videos in 2008 and 2012, which have also been watched by tens of millions of people. What made these videos go viral internationally was that Matt's dancing with different groups of people demonstrated the common humanity that unites everyone, from Massachusetts to Mongolia.

Online and offline, we all share, consume, and are influenced by unspoken forms of communication, such as pictures, memes, music, symbols, body language, and yes, dancing. The red equality banner in support of same-sex marriage had no words, but it was effective precisely because it spoke in a common visual language that was shareable and meme-able. In fact, the term "meme" was coined by Richard Dawkins to describe an idea that becomes powerful and contagious by social imitation and variation—sort of like a gene in the natural world, but for ideas.

We've also seen, recently, the emergence of a new universal language built around the hashtag. Hashtags are the words that follow the pound symbol you see everywhere online, but they're particularly popular on Twitter and Instagram. A hashtagged word serves as a kind of marker that leads you to other tweets or photos tagged with the same word. Hashtags can cross platforms and signify solidarity with an idea, whether it's on Facebook, Instagram, Twitter, or even written on a protest sign or spray-painted on a wall. Choose the right hashtag, and it can define a cause for the world.

Tips for Putting Tech—Life Balance to Work in Your Community

Today, change takes place on the go, in full view of society, and sometimes it can all get started with something as simple as a phrase that follows a pound sign.

One day the smartphone will shine its light on even the darkest corners of the world. When that happens, be there holding the phone, and use these tips to help do something amazing.

Focus on People, Not Likes

Lots of charities and groups are obsessed with attracting millions of followers and likes. Don't focus on the numbers. Remember that all change begins with a small group of committed individuals. Focus on adding value to people's lives. Post interesting and engaging content. The value of social media is not measured by just the number of people you can reach but also by the depth of the relationships you can build. Social networks allow us to find the core groups of people committed to change, who build the movements that eventually become millions strong.

Less Talk, More Action

The Internet can be a great way of raising awareness and sharing information about causes, but it's also a platform for collaboration. Give your supporters a set of actions or tangible activities they can participate in to show their support. Ask people to sign a petition, write a letter, or share a meme, or invite them to just donate to the cause. Tell them to film a YouTube video or tweet their thoughts, describing what your cause means to them. Change requires action. So, get people moving.

Tell Good Stories

Change happens when people are inspired. So, inspire them. The Internet is a wonderful platform for storytelling. Use all the tools of YouTube, Instagram, Facebook, Twitter, Tumblr, and blogging to explain what's at stake in your campaign, and humanize the issues. Everyone loves a good story. If you want to create change, you have to convince people of the need for change and show the tangible value in people's lives. A good story will convey all of that.

Use a Global Language

If you want to move the world, you need to speak the language of the world. This doesn't just mean having your words understood; it means sometimes you don't need words at all. Pictures, videos, music, or art often translate more easily across different communities and cultures by tapping into universal human values and emotions. If your movement can be defined by a simple image, a captivating video, or a snappy hashtag, these forms of communication can cross geographical borders and unite the world in action.

One of the most popular Facebook pages for employees, when I worked there, was the "Facebook Culinary Team" page. Facebook has a team of gourmet chefs who make delicious, free meals for employees—breakfast, lunch, and dinner—a wonderful perk that I sorely missed after I left and realized that my refrigerator was mostly just a storage unit for Diet Coke and Asher's bottles. The culinary team decided that rather than just hang a menu on the door of the cafeteria, they would create a Facebook page and use that page to post the day's culinary delights.

This quickly became a multiple-times-a-day must-visit page for all of us. Because the menus were only posted a few minutes before mealtime, we would all sit there, refreshing the page over and over again. It was imperative to know what was being served, so that if it was pasta day or taco day, you knew to drop everything and rush over immediately, or risk being caught in a thirty-minute line.

One day I realized something. There were about a thousand employees in Facebook's Palo Alto office at the time . . . but there were about four thousand people who had liked the "Facebook Culinary Team" page. That meant that there were three thousand people who had absolutely nothing to do with working at Facebook but just wanted to see what we were eating for lunch every day.

To this day, that page remains my creative inspiration. By speaking the universal language of food photos, the Facebook chefs were able to create an experience that was truly engaging and far-reaching.

Speaking of the power of photos, I was recently talking with the founder of an amazing SF Bay Area meal delivery service called Munchery. They send out daily e-mails with images of the food available for delivery that evening, and he told me that those e-mails get incredibly high open rates, much higher than the industry average for e-mail newsletters. When he surveyed some of his customers, they said they view the newsletter as "food porn" and look forward to opening the menus every day, even if they are out of town and not planning to order anything that evening.

Knowing how to reach people in an emotional way, and by using a universal language, can do more than just create a viral sensation or make people hungry. It might even mean the difference between war and peace. Take, for example the story of Ronny Edry.

In March 2012 an Israeli graphic designer named Ronny Edry, responding to the talk of war between Israel and Iran, posted on his Facebook wall an image of himself, holding his young daughter and the Israeli flag, below text that said, "Iranians, we will never bomb your country. We ❤ you." Within days, this image was liked and shared by thousands of people, including Iranians, some of whom meme'd their own posters, one of which read, "My Israeli friends. I don't hate you. I don't want war. Love ❤ Peace."

Suddenly, for the first time, enabled by social networking, Israeli and Iranian people were talking directly to one another, and to the world. Almost by accident, Ronny Edry started a peace movement, led by individuals and spread by likes and shares on Facebook and social media.

Perhaps something as simple as a message of peace and love will prove insufficient to stop wars, but as Ronny wrote on one of his later posters, "Making peace is a simple process that starts with each and every one of us. Every time we are sending a heart, it's another brick in the bridge we are building. Send a heart = make peace."

When we want to change the world, we don't need to look very far. There are movements everywhere crying out for the world's attention. If we speak the language of the world, we can work together to improve the lives of people everywhere, faster and with greater impact than ever before.

#letsgochangetheworld

 FUTURE

Everyone's a Media Company Now

It was the summer of 2011. I was standing in a checkout aisle of a Silicon Valley Target and getting in a quick round of Angry Birds. The green pig was teetering on the edge of the ledge, about to fall . . . any second now . . .

"Darn it!"

As I prepared to fire another bird, I was interrupted by a phone call that was slightly more important—and definitely the least likely phone conversation to have in a checkout line. It was my friend Andrew Morse, the senior producer at ABC News. He and I had worked together for the better part of eighteen months on our election-related collaboration at Facebook, back in 2007 and 2008 and then again on the midterm elections in 2010.

"Randi, congratulations! We've been nominated for an Emmy!"

I didn't believe it. In fact, I was so surprised that I could only stammer out a brief "Thanks! Congrats to you too!" before I hung up. And it wasn't until I got home and furiously Googled my name that I saw I had indeed been included in the team of correspondents nominated by the National Academy of Television Arts and Sciences, for Outstanding Live Coverage of a Current News Story—Long Form, for our work during the ABC-Facebook *Election 2010* coverage.

I suppose I had always assumed that if I was ever nominated for this sort of thing, I would find out when a man in a powdered wig would alight from a carriage, ring my doorbell, and with his white gloves, lift up one of those silver platter covers and present me with a calligraphy-inscribed envelope requesting my presence at the awards. But a phone call in Target was just as memorable, I suppose.

Plus, I got to say, entirely truthfully, that it really was an honor just to be nominated.

The ABC-Facebook 2010 election-night coverage played a key role in shaping my understanding of the increasing convergence between technology and the media.

During the ABC broadcast, I moderated a digital town hall at Arizona State University. My role was to integrate the traditional television coverage with the discussions taking place simultaneously on the U.S. Politics app on Facebook. The topics were a combination of serious news and crowdsourced discussions, ranging from tax cuts to war to the legalization of marijuana. While Diane Sawyer was the main election-night television anchor, ABC News' David Muir and I served as online correspondents, and the three of us engaged back and forth with one another over the course of the evening.

It was very exciting and very new. Working on a project that took advantage of both Facebook and the awesome power of television cameras gave me an incredible feeling of being at the epicenter of a new way of experiencing information and entertainment—a convergence between old and new media, and between television and the Internet.

I was quite literally seated at the intersection of television and the Internet, trying on one hand to talk to the television audience while at the same time interacting with the digital audience and feeding real-time insights to Diane Sawyer on-air. As it happened, I truly did have a front-seat view of a revolution in broadcasting and entertainment that forever altered people's ability to hear and be heard.

As recently as ten years ago, television and movie studios had a monopoly on how content was produced and distributed. Even amateur home movies of people hurting themselves were distributed to the masses only through televised platforms like *America's Funniest Home Videos*.

Though it seems hard to recall now, it wasn't that long ago when we were struggling to watch videos via a dial-up modem, record clips with terrible and complex camcorders, and upload our creations online. Sure, if you were using 3.5-inch disks to upload fifteen-second clips from your digital Sony Mavica onto Broadcast .com in 1999, you were ahead of the curve, but you were also the exception.

Now the ability to broadcast to everyone is universal, and the wildest dreams of the craziest futurists from the 1990s are a reality. Today, everyone is a journalist, everyone is an art gallery, everyone is a newspaper, a magazine, and a wire service, all in one.

And everyone is a mini media empire.

Once, if you wanted a mass audience for anything, whether it was art in a gallery or a show on television, you first had to get the approval of a few content gatekeepers. There was, after all, only so much gallery space or bandwidth to go around. And so, if your work was going to be shown, it had to be mostly profitable.

This is no longer the case. As long as you have a smartphone with a data connection, the whole world can be your audience. Thanks to the Internet, there are no more gatekeepers. There are no more limits on the human imagination.

This means that the artistic "scene" is no longer confined to the cultural enclaves of the big cities. Global cultural touchstones can and will come from everywhere—from the Harlem Shake to Gangnam Style, from Keyboard Cat to Grumpy Cat, from Nyan Cat to Lil Bub, and to the goats that yell like people.

This also means that a single person with a Twitter account and a good vantage point can give better on-the-scene news reports than the professional reporters. For example, during the recent tragic events at the Boston Marathon, Twitter was the go-to site for facts and information about the ongoing state of affairs. People no longer needed the news media to provide "coverage." They could go straight to the primary sources. At the time of the bombing suspect's ultimate capture, some 250,000 people were tuned in live to a simple webcam pointed at a Boston police scanner, which was being broadcast over the site Ustream.tv. Throughout the course of the chase, some 2.5 million people listened to this one scanner. Those are some serious ratings for a radio.

Under the right set of circumstances, and with enough talent, a random guy with an iPhone can be more influential than mainstream TV news. This is not to discount the tremendous talents who work on these shows. Nor is it the case that these shows won't still be important. It's just to acknowledge a truth already widespread: that the barriers to distribution have fallen, and it's open season on people's attention.

Of course, there's naturally a downside to the democratization of broadcast media and the de facto elimination of gatekeepers. Even if everyone is a media company, not everybody abides by broadcast standards and practices.

Because we're all so obsessed with "breaking news" and entertaining our followers within our own networks, with being perceived as the ones in the know and the first to the information—so much so that we now place more value on being fast than on being accurate—we put less value on being thoughtful, having an intelligent opinion, or taking other people's feelings and potential consequences into account. The pace of journalism has picked up so much that narratives are giving way to facts and, in the process, sacrificing understanding.

There are dangers to living life within a tweet's 140-character limit. Just because your average, everyday citizen *can* serve as a news source, that doesn't mean he or she should. Reputations of people, businesses, governments, artists, and ideas can all be built up or torn down in the blink of an eye or the push of a button.

The widespread dissemination of inaccurate or misunderstood information could have devastating real-life consequences. Wall Street traders rely on trading algorithms that read tweets for news about the world, and they automatically execute trades based on bad news. Also, people may be wrongly accused of crimes by the hive mind.

Take, for example, Sunil Tripathi, a Brown University student who went missing in the weeks prior to the Boston Marathon bombing in April 2013. When the manhunt for the bombing suspects was just getting started, the website Reddit—a popular and anonymous bulletin board—"up-voted" to their front page a story identifying Sunil as the main suspect. There was even a congratulatory post titled "Reddit gets it right," lauding the site's many users for correctly speculating that the bomber was the missing student. Within a few hours, the actual suspects were named, and the manhunt for them continued in earnest, but not before the

Facebook page set up to help find Sunil was bombarded with in-flammatory and painful statements from anonymous strangers. A few days later, the owners of Reddit issued a public apology, but by then the damage was done.

Giving everyone a megaphone tends to create a society that favors the loud and self-absorbed. Just because a lot of people are talking all at once doesn't mean anything valuable is being said.

In a world where everyone thinks of themselves as a leader, and everyone is shouting at their "followers," is anyone really listening?

The entire notion of celebrity has changed, but it hasn't neces-sarily been for the better. Your average citizens can go from being anonymous nobodies to being insta-famous, and sometimes with-out their consent or knowledge. Twitter makes it easier to be nice to established celebrities, but it's also easier to be really mean to them.

Everyone everywhere, no matter what they're doing, will stop in their tracks to watch a fight. It's almost an intrinsic human instinct. So, when someone uses the high platform of the Internet to tear someone else down, it'll collect hits for the sake of hits, but that may be all.

How do we navigate this new media environment? Who can we trust to get information both fast and right?

For starters, although certain websites that favor anonymity may have a hard time embracing the concept of authentic identity online, they should at a minimum embrace the notion that these are real people, whose real lives can be seriously affected by anonymous comments, and therefore try to act accordingly. When the mob is anonymous, it's also ephemeral and can vanish into the shadows of a deleted account when everything goes terribly wrong. Mobs have always sheltered and enabled bad behavior by way of the protection provided by anonymity. Online mobs are no different.

As for breaking news, it's probably better to review a variety of sources on Twitter and wait for a narrative to emerge from the gathering facts before jumping to conclusions. Speed doesn't equal quality, and we shouldn't believe everything we hear or retweet everything we see tweeted.

Finally, embracing authentic identity online means that, if you would not hurl abuse at someone's face in person, don't do it online either.

By the same token, don't be afraid to participate in the discussion taking place online. Your opinions are as valuable as anyone else's, and all the more valuable to your friends. The convergence of tech and media plus the elimination of traditional barriers to mass communication, while certainly complicating the media scene, have also made it possible to participate in the discussion in a way never really possible before.

As I sat in the correspondent's chair during the ABC-Facebook *Election 2010* broadcast, I had the chance to communicate to the television audience the thoughts of members of that audience. Like any good moderator, I chose the best things to say, as I saw them. Your voice might as well be one of them.

As it happened, the ABC-Facebook team didn't win the Emmy that year. That honor went to Anderson Cooper, reporting from a ditch in hurricane-ravaged Haiti, looking flawless in his tight gray T-shirt, even while surrounded by disaster. We didn't stand a chance. But it's all good.

Tech and Pop Culture

Before our *Election 2010* coverage, ABC News and Facebook had teamed up in 2009 to do some digital coverage at the South by

Southwest Interactive Festival in Austin, Texas. We were just wrapping up after a four-hour broadcast from the Facebook party-developer garage when the publicist for a tall, bouffant-ish British fellow named Russell Brand abruptly appeared and asked us to interview his client. Nobody had ever heard of the guy, but he had a movie coming out called *Forgetting Sarah Marshall*. I agreed to do the interview.

Things quickly went right off the rails.

No sooner had I mentioned Facebook than he replied he had a better idea for a social network called Cockbook. I'm not sure if he meant a social network dedicated to junk in general or just his. Needless to say, the interview ended pretty quickly.

I decided not to propose the idea to the Facebook board. But I was happy to be doing interviews and a little surprised at the fact that, without a television network to call my own, I had nevertheless ended up as an entertainment correspondent. This is all thanks to the Internet.

Technological revolutions have always produced new kinds of public figures. There were no movie stars before Thomas Edison invented the movie camera. Technological progress has always influenced what people read, watch, and listen to. And pop culture is being defined more than almost anything by technology and the Internet.

Back in the early days of Facebook, one of my first tasks in consumer marketing was to manage how our company was mentioned in movies, TV, and print media. I estimated that the kind of free marketing we could get was worth somewhere in the range of hundreds of millions of dollars. Nevertheless, we turned a lot of it down.

The first shows that wanted to use our branding were mostly crime dramas, which would associate Facebook with grisly murders or stalkers. As lucrative as the placements might have been, it was very important to us that our brand not be perceived in this context, and we turned down all these offers. The shows would inevitably go with some rip-off site, such as MyFace or Facester.

There were many exciting opportunities we said yes to, though, and before we knew it, Facebook was at the center of pop culture. That's when the celebrities came calling.

Back in the mid-2000s, it was far from clear that Facebook was going to become the dominant social network in the world. We had quickly cleared Friendster, but we were still second to MySpace, which was ferocious about attracting celebrities to its platform. Kids of that era were still being called the "MySpace generation."

So, with my colleagues at Facebook—Dave Morin, who is the current CEO of Path, and Chris Pan—we assembled a crack, ad hoc A-team to get celebrities to use Facebook. We made it our mission, on our own time, to be Facebook evangelists to the celebrity community. This was how I ended up spending a day at Ashton Kutcher's house, explaining what a Poke was (answer: we still don't know), and found myself in the basement beneath a Britney Spears concert, conceiving a merchandise line of classic Britney costumes that would be sold on the site as virtual gifts for charity.

There was a lot of internal debate within Facebook about whether we should be focused on celebrities or not. Some people in the company thought working with celebrities was a waste of time. Others felt celebrities were a big driver of cultural influence and should definitely be a priority. Others just thought Ashton Kutcher was cute and wanted him to visit Facebook; they didn't really care if he joined the site or not.

As much as we wanted to change the world by connecting its people, we were still running a company, and celebrity endorsements were incredibly valuable promotion and marketing. Not only that, but because of what Facebook and social media were doing to pop culture, celebrities were already starting to come to us.

We took a bit of a stealthy backdoor approach. And then, in 2009, we watched, rapt and jealous, as Ashton Kutcher and CNN got into a literal popularity contest to see who would be the first to receive a million followers on Twitter, a novel new social network and rival. That race pretty much catapulted Twitter's entire business. Nobody questioned our little celebrity squad after that, and we were encouraged to sign big names.

Even though he played a dopey guy on television, Ashton Kutcher was a visionary when it came to understanding the power of social media as a broadcast medium, on par—or potentially on par—with the television networks of the day. In 2009, he was quoted as saying that he found it astonishing that one person on Twitter could have as large a voice as an entire media company, and that if he beat CNN to one million followers, he would ding-dong ditch Ted Turner.

So, not only did Ted Turner get his bell rung, Ashton now has over fourteen million followers on Twitter. Broadcast power like this is historically unprecedented. The guy who played Kelso on *That '70s Show* has a broader platform, and direct access to more people, instantaneously, than any newspaper in the pre-Internet age.

It wasn't long after Ashton set the bar that celebrities came streaming into Facebook's office by the dozens, always tailed by very nervous-looking publicists. It was during that era of incredible growth at Facebook that Kanye West jumped up on a table in our cafeteria to freestyle some lyrics. We later took a photo with Kanye in front of one of our conference rooms, which we had lovingly renamed the "Imma Let You Finish" room, after he infamously interrupted Taylor Swift's acceptance speech with that line at the *MTV*

Video Music Awards. When *SNL* star Andy Samberg came to visit Facebook HQ, he dressed like Mark for a day and confused a few engineers. And at one point, I was about to ask the random dude sitting at my desk to go away, when I realized it was Keith Urban. Then I had to tell Keith Urban to find another place to sit.

Every celebrity who visited our headquarters or took advantage of our platform understood that social media had completely changed pop culture. The concept of a polished, packaged celebrity image was no longer available. People expect you to have a social media presence, and to have access to you via it.

Sure, people could always have joined celebrity fan clubs. But, beyond the fact that fan clubs always feel a little weird, social media is a different beast entirely.

Moreover, the definition of "content" was changing. Just as "news" was going from something Walter Cronkite once told you to something your friends shared with you, "content" was going from something professionally produced to a five-second clip recorded on a smartphone or a passing thought tweeted while on the can.

Of course, not everyone with an iPhone is an immediate media mogul, a kind of mini Rupert Murdoch with a data plan. But the point is things are complicated. The social-content sphere is important, but so are the movie studios that still produce movies and the TV studios that still produce TV. Instead of the Internet surpassing traditional media platforms, it's being incorporated into it, along with all of its quirks and challenges. And the content producers and celebrities need to adapt. Because clearly most people already have.

Back when I was in high school at Horace Mann, I was something of an outcast for being a "geek," which is now cool. School kids in America and around the world are now almost as keen to be the next Steve Jobs as the next Michael Jordan. It really is incredible.

One mom, according to the Internet, named her baby "Hashtag," an Israeli couple named their daughter "Like," and an Egyptian couple named their newborn "Facebook."

The most popular movies and television shows now have plotlines that revolve around technology and the people who use it. From *The Big Bang Theory* to *The Social Network* to *Shark Tank,* nearly everything that shows up on television today has a tech-focused element to it. Plus, you can't watch a TV show, a commercial, or even a viral video on the Internet without seeing a hashtag, Facebook URL, or Shazam logo.

Pop stars are angel investors. At last count, the ranks of celebrity venture capitalists included not just Ashton Kutcher, but also Lady Gaga, Jay-Z, Justin Bieber, Justin Timberlake, Britney Spears, and Kim Kardashian. It's not rare anymore to show up at a tech-centric event such as Y Combinator Demo Day (where new start-ups are demo'd for the first time) and see Ashton, M. C. Hammer, and will.i.am speaking on a panel or sitting in the audience. Hardly a day goes by without a celebrity paying a visit to the Twitter office to try the bacon plate or stopping by the Googleplex to play on the beach volleyball court.

Tech is the new pop culture. Geek is the new rock star. And the tech community should leverage its pop culture cachet for good.

I recently participated in a small roundtable discussion with White House Senior Advisor Valerie Jarrett about the best way to get girls and women interested in learning to code. Silicon Valley's culture is often dominated by men, so the tech community really needed to find a way to increase the participation of women in this area. The guys on the panel seemed to think that the way to go about it was to make coding a mandatory subject in schools. I disagreed and suggested that the best way to get girls to code was to make it cool in pop culture and the media.

This isn't to say that girls will only do something because it's cool or sexy, or that if you want women to get involved in technology, all you really have to do is sparkle up a few MacBooks. (Though, if you want to sparkle a MacBook, by all means sparkle away.) It means that both the technology and entertainment industries have a role to play in providing positive, tech-centric role models for girls. Make coding cool, show how it changes people's lives, and the kids will code.

This was why, when MTV came to us and wanted to do *Diary of Facebook,* a behind-the-scenes special on the company, I insisted that we not only feature our engineers, but also bring in people from the outside world whose lives had been changed by Facebook and have them meet the programmers on camera. By connecting the human and technical dimensions of technology, we could help show the tangible impact that technology has in the lives of real people. So, we brought in an Army veteran who, while deployed, was able to see his son being born via Facebook video chat and had him meet the engineer who had built that application. We did the same for people who managed the adoption of their children via Facebook messages, along with a few other amazing examples.

When the real-world impact of a job is clearly demonstrated, people feel far more invested in the work. As Adam Grant demonstrated in his recent book, *Give and Take,* when students who worked in a university call center were exposed, even for ten minutes, to the stories of the individuals whose scholarships their work helped fund, these students were on the phone with alumni 142 percent longer and increased donations by 171 percent.

Telling the story behind the work is essential to humanizing it and making it relevant and important. When pop culture can do this for the world's coders, programmers, and underappreciated sysadmins (systems administrators), we will see more women, and more everyone, working in technology to foster change in the world.

Content Matters

Every year or two, Facebook hosts a social media developers' conference known as F8. At these annual events, Facebook announces new developments and invites anyone who is building their business on top of the Facebook platform to meet the team and hear about announcements firsthand.

About a week prior to the April 2010 conference in San Francisco, the volcano Eyjafjallajökull erupted in Iceland, sending into the atmosphere an ash cloud that grounded almost all air traffic in Europe. About a fifth of the people who were supposed to attend the conference could no longer come.

Facebook went to work. We had to solve this problem. It was obvious that we could try to live-stream the keynotes, but I didn't think that was nearly enough. Instead, I set out to create a mini media company online. Working with our engineers, I helped create "F8 Live," a platform that would broadcast on three channels at the same time, three different feeds from around the conference, including all the keynote sessions and a series of hosted interviews with developers. This way, people stranded by Eyjafjallajökull would feel as if they were actually attending the conference. We even made a point of filming the breakout sessions and taking gratuitous pictures of the food.

At the end of the day, a hundred and fifty thousand people tuned in to the conference online, which is really pretty incredible when you consider that it was a super-technical conference and there were far fewer than a hundred and fifty thousand people who had been stranded and unable to attend.

This experience got me thinking. We were starting to innovate new ways of making technology and media work together. But why were we doing this only during the developer conferences? Why weren't we doing something like this all year round, like a kind of Facebook TV channel?

What would a TV channel look like, built entirely on top of Facebook?

A few months later, at a late summer all-night Facebook hackathon, I got a chance to test my idea. Instead of "F8 Live," what if we called it "Facebook Live"? We had so many interesting people, celebrities, top executives, and politicians constantly coming through the Facebook office—why not put some of those visits to use on our very own Facebook broadcast channel?

The following week, the president of the Sports Illustrated Group came to visit Facebook, and I interviewed him, asking for his ideas about the future of sports, tech, and media. It was an interesting conversation, and I wanted to do more.

Unfortunately, there weren't a lot of resources available for this project. So, I duly commandeered what looked like a broom closet, fixed a sign to the door proclaiming "Facebook Live," and went to work securing interviews.

I had a lot of fun in the ensuing months. While "Facebook Live" was still only a very small passion project, alongside a very busy day job leading the consumer marketing team, it nonetheless started to pick up some very real traction.

America Ferrara came to promote her new indie film.

Then, one day, Katy Perry's manager, Glenn Miller, called, asking if Katy could come on "Facebook Live" to announce her new concert tour. Katy had gotten her start online, and they had been searching for an innovative new way to launch her tour. So, in January 2011, Katy Perry showed up in a pale-blue lace dress and sky-high heels for a tour of the campus, a taste of Nacho Wednesday, and a "Facebook Live" broadcast with yours truly.

Our online conversation was a lot of fun. Katy answered tons of fan questions that streamed in on my laptop during the interview, and she announced the North American leg of her California Dreams 2011 world tour on Facebook, before anywhere else. Her tour sold out within minutes, and her "Facebook Live" appearance generated a lot of press attention.

Once the Katy Perry interview happened, "Facebook Live" was official and there was no turning back. Mike Tyson was a lovely gentleman, who told me all about his beloved pigeons. Pee-Wee Herman inspired mobs, clamoring to get a glimpse. Linkin Park band members wore full-body Facebook logo costumes. Kanye West played his new single at top volume and talked about the origins of his diamond teeth. One day it was Texas governor and presidential candidate Rick Perry, and the next it was Conan O'Brien.

Snoop Dogg was supposed to do an interview once, but he slept through it, for hazy reasons. He later posted a video, in which he said to me, "Randizzle, when I come back to the Bay, we have got to play." I've got to say, that was quite a day.

Things were getting crazy. I was a long way from Dobbs Ferry. And yet I knew there was further to go.

I leveraged my experience as the host of "Facebook Live" to secure a spot on the famous red carpet at the Golden Globes, as their official online correspondent. Prior to the event, I sourced some questions from the show's Facebook audience that would be asked, live, of the attendees and broadcast over Facebook. As usual, the Internet audience came through with some interesting questions. When Paul McCartney passed, instead of asking him the standard questions about his outfit, I yelled out, "Paul! What's your high score on Beatles Rock Band?!" He laughed and said it was definitely the most creative question he had been asked all day.

Later, when James Cameron passed, his eyes lit up when I asked him to explain to the Internet viewers some of the tech specs behind the new cameras he had invented to make *Avatar.* I felt particularly ready for this moment, since I had spent the previous day interviewing a fake "James Cameron" while media training with the talented CEO of Clarity Media Group, Bill McGowan.

Later on, I got to walk the red carpet myself and by coincidence was trailing Dana Brunetti, producer of *The Social Network,* which had come out that year. I thought it would be pretty funny if I could get a picture with my brother's movie alter ego. But before I could make that happen, the cast came up to me and asked to have their photo taken with me!

It was all happening, and it continued right up until that fateful day when Obama came to Facebook, using my little hackathon project, "Facebook Live," to speak to all of America.

My little broom closet media company taught me some very important lessons about the future of media. The most important lesson was that knowing how to navigate and use social media was to be its own special kind of expertise and an essential one for journalists and broadcasters in the future. I also saw that the online etiquette around journalism was changing rapidly. In a world where people can finally yell back at the television and be heard, those on the screen better be prepared to listen. It's no longer enough to be a pretty face reading lines off a teleprompter. In the new connected world, the teleprompter talks back at you. The media correspondent of the future will need to have a new kind of skill set: the ability to be a correspondent, a community manager, a curator, and a member of the audience, all at the same time.

This will not be an easy task to accomplish. It will be a very difficult transition from everything the television industry is currently used to. Anchors will have to anchor more than just the one newsroom they're sitting in and become proficient at monitoring a social media stream for commentary on the proceedings. Media companies will have to start fostering a bench of talent who can do this.

People watching television news no longer want just a well-groomed face reading the day's headlines from cue cards. They want someone who is part of the dialogue they also have access to all over the Internet. This needs to be a person who is also willing to go off camera and engage with his or her followers, someone who cuts above the clutter with his or her own point of view.

Since the rise of social media has made the crews that run television much leaner than they used to be, this anchor of the future will need to have some video-editing and production experience. The budgets are smaller, and the time frames are shorter. With a skeleton crew, that anchor will need to get everything done in a tenth of the time.

The Internet has given birth to a whole new middle layer of content, people who want to be heard for five minutes on a topic they're passionate about and may not have much else to say beyond that. If before there were only the professional broadcasters and everyone else, there is now that middle layer—the semi-pro bloviators, if you will—thousands of people who have something very valuable to say on a single topic, but not on every topic. These are the microprofessionals and opinion proclaimers that populate aggregation websites like *The Huffington Post*. These are the "guest posters" on your favorite sites across the Internet. Entire media companies are being built on the backs of people who have something interesting or provocative to say in five to eight hundred words and then immediately retreat back into obscurity.

There is an overflow of companies in Silicon Valley all threatening to "disrupt" television, all cluttering the experience of watching TV with an array of apps that are more third rate than second screen. Television doesn't need to be disrupted by Silicon Valley; it needs to be embraced and enhanced.

We have entered this interesting new world where every person and brand needs to think of itself as a media company, as a content creator with a sharing strategy. For example, Starbucks will spend millions of dollars on advertisements to potentially reach their customers on morning TV shows, such as *Morning Joe*. But Starbucks has so many followers on Facebook alone that any message they broadcast on the Facebook platform will reach more people than a similar advertisement on every single morning show combined.

An online platform gives companies direct access to millions more accessible, demographically identifiable individuals following them on Facebook, Twitter, and other social networks. There's no reason that Starbucks or JCPenney or Macy's or any number of other companies shouldn't make their own morning shows and broadcast to their followers on the Internet.

In the early days of television, brands directly sponsored shows all the time, and the price of doing so was tremendously high. Now, the cost of bandwidth has fallen so low that, if a brand were willing and had the capacity to produce some kind of original content, that brand could get a tremendous bang for its buck by going directly to the audience. Heck, the "soap opera" was born out of product placement.

Social media provide brands with a good deal of specific information about their audience. There's no reason brands couldn't leverage that information to produce better content for their audiences. Would people watch the "Starbucks Coffee Morning Hour" on their Facebook page? Why not? Especially if it was entertaining and did the work that a lot of TV morning shows need to do, which is be informative without being too overbearing.

And it's not just brands that must be their own media companies. The same holds true for individual artists. Just as everyone has the potential to be his or her own publicist, celebrities also have to be their own publicists, promoters, directors, and ticket agents. Recently, the comedian Louis C. K. made headlines by skirting the standard distribution channels by selling his own tickets and distributing his material online himself.

Not only that, but celebrities can also tap into the Internet to fund their projects in ways that were never possible. After an entire *Veronica Mars* movie was funded through donations on Kickstarter, Zach Braff was able to crowdfund a film in only three days.

The new producers can be everyday people. And the people want to be heard.

When I got the chance to create my own new media company, I took note of what I had learned from my experiences at "Facebook Live" and set out to change the way media and pop culture interacted with each other. All it took was an Icelandic volcano, a small broom closet, and a dream.

TV Still Matters

One of my first goals when I left Facebook was to produce a television show featuring the lives of people in Silicon Valley. I knew that, notwithstanding the rise of Internet-based media, TV still mattered a lot and was the driving force behind pop culture. Even though I was creating an Internet-based media company, I couldn't just upload a few videos to YouTube and call it a day. If I had done that, nobody would have paid any attention to the show. There were still Emmys to win, more audiences to capture, and more work to be done on television.

Even as the distinction between TV and the Internet lessens, TV is still the dominant force driving the narrative, curating the content, and shaping the culture. In 2013, the Pew Research Center's annual *State of the News Media* report showed that when asked where they got their news "yesterday," over 50 percent of Americans said via television, while only 39 percent said via online or a mobile device.

I know I'm expected to say, "Forget Hollywood. Digital is coming to eat your lunch!" But I don't think that's true. I think New York, Los Angeles, and Silicon Valley are going to share the meal, and there's enough food to go around. Even YouTube puts out a press release every time a show that began on their platform makes the leap to TV. That, in and of itself, is evidence that TV still matters. If at one time TV was the only source of content, at least now it also serves as a curator.

There is still the common assumption that, if something was really good, it'd be on TV. Maybe that will change in an era of Netflix-produced content, but it hasn't changed yet. *House of Cards,* Netflix's in-house drama starring Kevin Spacey, cost $100 million to produce and was truly a breakthrough moment for premium content online. In the meantime, getting a show on TV was a way for my company to get the stamp of approval from the status quo powers that be and break through the noise.

Even though I had come from Facebook, I found that people in Hollywood weren't taking me as seriously as I had hoped. They saw me as "one of those Internet video people," and one person even commented that "it's cute what you guys are doing down there," even though last time I checked San Francisco was north of Los Angeles. It was hard to get the appropriate meetings, but I was determined to prove them wrong. It was always a dream of mine to produce a television show, even though it seemed like a dream that had a very small chance of coming true.

Who would have thought that the opportunity would come to me via reality TV?

In the fall of 2011, I found out through the grapevine (and some very diligent work by my colleague Erin Kanaley) that Bravo was casting a Silicon Valley reality show. We reached out to get more information. Bravo responded enthusiastically, offering me a part on camera, in the cast of the show. I thought about it for a hot minute and then declined. I may have lost the Emmy to Anderson Cooper, but I was not ready to become a reality TV star.

Bravo came back with an interesting proposal. They had been looking for someone in Silicon Valley to sign on as executive producer—would I be interested in that? The more I thought about it, the more I saw that this show could be the foot in the door to Hollywood that I had been looking for. After all, it's not every day that, if your goal is to create a television show, one just suddenly appears out of nowhere and falls into your lap.

A few months later, Bravo announced *Start-Ups: Silicon Valley*, a reality show that followed the experiences of six entrepreneurs and shined a light on tech culture, and my participation as executive producer. Soon after inking the deal, we went into production. In true start-up form, I learned as we went along.

As with the best of the genre, our reality show had deep interpersonal drama, hard workers who also threw legendary parties, and your standard mix of fun, craziness, money, and love, all jammed into this zany community called Silicon Valley.

We had a huge amount of buzz leading up to the premiere. This was the year of the Facebook IPO. Tech accelerators and incubators were popping up all over the country, and talk was in the air about "bubbles" and "series A crunches." Building on this zeitgeist, it made perfect sense to combine the fun, watchable drama of reality TV with the nation's growing fascination with the technology community.

Not all the buzz was positive. In fact, there was plenty of pushback right away.

After working at Facebook, my decision to work in the "old media" of television seemed like a betrayal to the entirety of "new media." Some tech bloggers took me out to the woodshed for daring to make a reality show about Silicon Valley that showed pretty people having a good time, instead of the "reality" of Silicon Valley—presumably a guy sitting in an open-office layout, typing on a laptop for ten hours, then passing out in a luxury shuttle van and drooling all over his hoodie.

I guess the fact that the show aired on Bravo wasn't a strong enough hint that this wasn't intended to be a gritty CNBC-style documentary. Perhaps these same writers would criticize the *Real Housewives of New Jersey* for failing to accurately portray the difficulty of getting grime off shower tiles and making a nice roast before the husband comes home.

Anyway, whether the buzz was positive, negative, comedic, or downright hateful, everyone was talking about it. When we became a trending topic on Twitter the night before the 2012 presidential election, I was ecstatic.

Unfortunately, we did not see a correlation between the insane buzz and our actual ratings, which started high but then dropped and remained disappointingly low.

I think the problem with our show was that it was too techie for the reality crowd and too much reality for the tech crowd. Just as venture capitalists caution start-ups to start small, delight their initial users, and grow, we would have benefited from picking one audience and delighting it. In hindsight, I think we needed to do a show that either targeted people with very little knowledge of tech, and used Silicon Valley as just a backdrop, or targeted people interested in tech entrepreneurship, and went much deeper into start-up life, filming more in start-up offices, showing cofounder dynamics, the pressures of competition, the perks, and so on. We took a middle ground, trying to appeal to both audiences and wound up not delighting anyone.

Regardless of the final numbers, I know *Start-Ups: Silicon Valley* had an impact. We changed the lives of our six entrepreneurs, we inspired conversations in Hollywood about more Silicon Valley–based TV shows, and I personally have heard from thousands of people who, after seeing our show, decided to pursue their dreams and either study computer science, return to grad school for entrepreneurship, take up coding, or apply to a tech incubator.

When I first got involved with this show, I said that if I could inspire even one woman in the middle of the country to pursue a career in tech, I would consider it a win. It was a win.

I considered it also a personal victory in a few ways. After that show, I had the television broadcast street cred to take more serious meetings in Hollywood. It's a big deal to get a television show that becomes a series and an even bigger deal if you're a first-time producer.

Plus, I had done something big and loud and proud that was all my own.

I remember when my colleague Bradley forwarded me a review in which the show had been awarded zero out of five stars. Rather than get angry or disappointed, I started laughing and jumping for joy. But it didn't have anything to do with the show. It was the very first snippet about me I had ever read that didn't say "Randi Zuckerberg, Mark Zuckerberg's sister." Instead, it just said "Randi Zuckerberg, television producer." To me, that was worth six stars out of five.

I founded Zuckerberg Media on the premise that tech and media needed to work together. San Francisco and L.A. needed to team up, not fight each other. Both worlds have the same goals: delighting users, capturing their attention, and creating brand loyalty for recurring engagement.

But these worlds are also extremely different. Part of the excitement for me is in figuring out how to bridge the two.

It's tough being a bridge. Collectively, the people on my team have created thousands of hours of television, reached hundreds of millions of people with content on Facebook, Twitter, and YouTube, and worked with dozens of A-list talents, and we're still trying to prove ourselves. But we're getting there.

Like any start-up, you launch your product in "beta mode," listen to feedback, revise like hell, and eventually either do something awesome or run out of money trying. Working at Facebook taught me to move fast and break things, and that sometimes launching is better than being perfect. We launched this show quickly. We worked our butts off. We learned a lot in the process. And we'll nail it next time.

Whether we do it only online, only on TV, or most likely, a high-quality hybrid of the two, I'm not giving up, especially now. We are here at the vanguard of change in two global industries. And we're just getting started.

Tips for Achieving Tech—Life Balance in Communication

Better to Be Right than Quick

In a world of ubiquitous information, knowing something that other people don't carries with it a special kind of cachet, regardless of the quality of the "information." This is why whenever a celebrity passes away, people rush to the Internet merely to be the first to post about it, as if that mattered more than making a statement about their relationship to that person and what that person's life meant to them. There's no value in this kind of behavior, and sometimes—as in the aftermath of the Boston bombings—this rush to be first can lead to very damaging consequences for innocent bystanders. It's okay not to be first to say something and far better to be right.

Don't Be a Jerk

Being a jerk to other people via the Internet doesn't make you a rare, valuable truth teller. It just makes you a jerk.

New Skills for a New Time

Film, music, and television stars were once supported by large crews of roadies, aides, publicists, and managers. Now you're all of those things, on your own. Don't just learn to write; learn to upload your own content. Use the free educational tools offered online to increase your skill set, now that the demands for a social existence are so much greater than they ever were. If you haven't yet learned to tweet, blog, or upload photos to Instagram on your own, get on that.

.   .   .

The Internet is an incredible force for good. Vast crowds of helpful people can be mobilized online almost in the blink of an eye.

At the same time, you have to remember that anytime you share, you open yourself up to the judgment of others. Sharing is a wonderful thing. The positives far outweigh the negatives. But make sure you have your thick skin and your big-girl pants on, because people can sometimes be a bit cruel.

The more successful you are and the more you have to say, the more people will be mean to you on the Internet. The only way forward is to embrace your haters. Don't be afraid of the keyboard cowards. Engage them. I used to get nauseated at the slightest hint of a mean blog post or negative tweet. Now, I try to celebrate them.

Attention is a currency. I often tell any people or organizations that are terrified of negative online comments that "haters are an inch away from loving you." If someone is taking the time to read something and respond, sometimes all that person wants is to be heard. With effort, you might be able to turn that person into a passionate, local, enthusiastic advocate. And even if you can't, haters keep you in the conversation. Love me, hate me, but just don't forget me.

There will always be people who are jealous or afraid, who don't have the courage, strength, or conviction to make an impact on the world. Don't let their insecurities and frustrations with their own lives slow you down or affect your day. They are just words on a screen.

Part of tech–life balance is being able to put down the phone, close your laptop, and not take your digital baggage with you. Never believe the online hype. You're never as awesome as people make you feel online. And you're never as bad as people make you feel online. The only relationships that really, truly matter are those you have with the people right next to you when you stop text messaging and put your e-mail aside.

It's often been said that your Facebook profile isn't the real you, that it's just the best of you. Let's change that. Let's make the real you also the best of you. When you live an authentic life both online and off, strive to be honest. Strive to find personal peace, friendship, love, fulfillment at work, and good in your community, and use the Internet to improve your life, not control it.

Technology gives us the power to change the world. Let's start by changing ourselves. Let's make our complicated wired lives a little bit easier, and a lot more wonderful.

FINAL THOUGHTS

A year after I was asked to get M. C. Hammer a seat at President Obama's "Facebook Live" town hall, I found myself on a boat on a sunny day on the San Francisco Bay, headed to Oakland with M. C. Hammer himself and one of my investors, Jody Gessow. We were on our way to Jack London Square in Oakland to check out a potential production space for my new venture: Zuckerberg Media.

When going to Oakland, who better to bring along than hometown hero M. C. Hammer? And what better way to go than by boat?

The sun glinted off the choppy waters as we sailed beneath the Bay Bridge.

Hammer turned to me. "Randi," he said, "I've got an idea. We should create a production company together."

"I love it!" I said. "What do we call it?"

We both fell to musing.

Suddenly it came to me. "$Z = MC^2$!" I shouted.

Silence.

That was it. I had just officially out-nerded everyone on the boat. But then Hammer laughed.

As the boat sailed on to Oakland, I caught a glimpse of San Francisco behind me and had one of those terribly poignant personal moments when you sigh and remark to yourself about how strange life is and how quickly everything changes.

Not that long ago I was shoveling snow with my brother and sisters in a Dobbs Ferry cul-de-sac. It felt like just moments ago I was belting my heart out as Peggy in *42nd Street* at Horace Mann and "stopping to smell the flowers" in music class. I remembered singing "Zombie Jamboree" on stage with the Harvard Opportunes at my first concert, and then, in the blink of an eye—four years later—my last. I remembered the Naked Cowboys, Midtown drinks on warm summer Fridays, and getting that BrentT instant message. I remembered the plane ticket to California to work for The Facebook, a teary good-bye with Brent, followed by an even more teary reunion, a love affair, a marriage-by-Outlook invite, and a beautiful blond boy named Asher, whose happy photos fill my Facebook Timeline.

I took my phone out of my pocket and began to flip through photos of Asher, looking for a good one to post. There weren't any, so I started to photograph my view of San Francisco, but I stopped myself, put the phone away, and turned back to ask Hammer some more questions about Oakland, his career, and his life.

This was no time to be on the phone. There would be time for that later. The little five-ounce communication device I held in my hand was something directly out of *Star Trek*. Always-online smartphones and social media have completely changed the way everyone is interacting. The previously solid boundary lines between friends, lovers, families, work, society, and celebrities are all beginning to blur. Technology seems to be making things both easier and harder at the same time. We have at our disposal incredible communication devices, but we seem to forget how to communicate with one another.

For me, the way out of the mess is understanding something essential about the tools of technology, that they are exactly that:

tools—meant to make life better, not worse. And by living an authentic life online, we begin to understand how to use these tools to achieve a proper tech–life balance.

And so, like most stories go, it all comes back to Hammertime.

As M. C. Hammer once said, "You either work hard or you might as well quit." Well, I don't know about you, but I plan to keep working hard for as long as I can.

We will all have career highs and lows, money gained and lost. We will have tremendous victories and crippling losses. We will have baby photos, graduation photos, wedding photos, and then . . . baby photos all over again. We will have close friends who disappoint and strangers who delight. But isn't it that sharing, that vulnerability, that human connection that makes life so wonderfully worth living?

It's easy to hide behind a screen, a text message, a photo, an e-mail. The hard part is truly getting out there and living your life, being true to yourself and connecting with others. Technology has shown us a new world. But there's work to be done to make that world a beautiful, hospitable place for our generation and the generations to come.

Work hard. Play hard. Post hard. Tweet hard. But most important, live hard. Because we are way too legit to quit.

ACKNOWLEDGMENTS

Wow! You've made it this far! Thank you!

Thank goodness I have more than 140 characters, because there are so many people I am eternally grateful for and indebted to.

Thank you, first and foremost, to my incredible book team at HarperCollins. Mark Tauber, thank you for your perseverance and patience. Gideon Weil, thanks for being a true teammate and extraordinary editor. Lisa Sharkey, thank you for bringing so many creative ideas to the table that I've lost count. Margaret Anastas, thank you for embracing both me and Dot so wholeheartedly and lovingly. And thank you to Claudia Boutote, Laina Adler, Suzanne Quist, and Suzanne Wickham, for all the hard work that went into turning my dreams into an actual, physical, printed reality . . . and then making sure everyone knew about it!

Thank you to all the readers of my Dot Complicated newsletter and website. It is because of your support and loyalty that this book was even possible and that I wake up every morning passionate and excited to get to work.

Thank you to my former Facebook colleagues, especially my consumer marketing family: Alex Wu, Aubrey Sabala, Charles Porch, Elliot Schrage, Erin Kanaley-Famularo, Jonathan Ehrlich, Larry

Yu, Mandy Zibart, Matt Beaman, Matt Harnack, Matt Hicks, Skip Bronkie, and Tish Stenson. Special shout-outs to Brandee Barker, Meenal Balar, and Raquel DiSabatino, my soul sisters in marketing/comms.

Thank you to the wonderful mentors I've had along the way: Dan Rosensweig, Dana Brunetti, Francine Hardaway, Jason Goldberg, Kathy Kennedy, Leslie Blodgett, Mike Murphy, Ray Chambers, Ron Conway, Shari Redstone, Shervin Pishevar, Sheryl Sandberg, and Terry Semel. I have always valued your vision, your honesty, your leadership, and, when needed, your tough love.

Thank you, Andy Mitchell, for giving me one of my biggest career breaks to date. I love our jet-setting friendship, from Florida to Buenos Aires to Brooklyn.

Thank you, Andrew Morse, for being a partner in crime over and over again. One day, we'll win that Emmy together!

Thank you to the Academy—it was truly an honor to be nominated. ;-)

Thank you, Kevin Colleran, for making that very first Facebook holiday party video with me. I don't think I'll ever have as much fun filming anything as we had together. Thanks to everyone who was a good sport about those videos, especially Jeff Rothschild. We did it out of love.

Thank you, Ari Steinberg, Ezra Callahan, Luke Shepherd, Peter Deng, Tom Whitnah, and the all-star engineering team that made those initial amazing politics projects happen. Thank you to Tim Kendall for joining me on air. Thank you to Adam Conner, Andrew Noyes, Chris Kelly, Dan Rose, David Fisch, Denise Trindade, Ethan Beard, Julia Popowitz, and Matt Hicks for your support and friendship. Working with all of you was a true highlight of my career.

Thank you, David Prager, my parody-music-video-making soul brother. Don't you wish your cell phone was hot like me? ;-)

Thank you to my band, Feedbomb—Chris Pan, David Ebersman, Andy Barton, Bobby Johnson, Eric Zamore, Eric Giovanola, and Sean Chaffin—for continually giving a geeky soccer mom the experience of being a rock star. I adore performing with you and hope we can keep playing together for a long time to come, no matter how many times Chris Pan threatens to quit and move to L.A. Pre-Feedbomb, thank you to Evanescence Essence—Chris Kelly, Bobby Johnson, Chris Cox, and James Wang—for rocking out so hard. Chris Pan would yell at me if I didn't take this opportunity to say that Feedbomb is available to play at your next wedding or event or conference. ;-)

Thank you to Woody Howard at Horace Mann School for that life-changing role as Peggy Sawyer in *42nd Street*. You truly shaped so much of my life, and my online identity, influencing years of screen names and Internet handles!

Thank you to the Harvard Opportunes—for your beautiful music, and even more important, for the lifelong friendships. You guys are aca-mazing.

Thank you, Yossi Vardi and Michelle Barmazal, for the amazing honor of singing at Shabbat Dinner at Davos. And thank you to Matthias Luefkens, Diana El Azar, and Josette Sheeran for the amazing honor of even setting foot in Davos in the first place.

Thank you, Michelle Myers, Elise Wood, and Bob Sauerberg at Condé Nast and Andy Levey and Lou D'Angeli at Cirque du Soleil, for taking risks on zany ideas that turned out to be amazing.

Thank you to Lauren Zalaznick and Eli Lehrer at Bravo, and to Evan Prager, Jesse Ignjatovic, and the whole crew at Den of Thieves, for giving a TV newcomer her racing stripes.

Thank you to the journalists who have loved me, hated me, and heckled me throughout the past years—special thanks to Kara Swisher, Ken Yeung, Liz Gannes, and Owen Thomas for always encouraging me to belt my heart out.

Thank you to so many amazing "wired" women who have helped and inspired me throughout my journey: Abby Ross, Ali Pincus, Ann Brady, Anne Fulenwider, Angelina Haole, Arianna Huffington, Brit Morin, Cathy Brooks, Desiree Gruber, Elizabeth Weil, Farzana Farzam, Hasti Kashfia, Hillary Frank, Jennifer Aaker, Jennifer Lima, Jessica Melore, Johanna Argan, Julia Allison, Julia Popowitz, Julie Vaughn, Kara Goldin, Katherine Barr, Kirsten Green, Lea Goldman, Libby Leffler, Melissa Sobel, Porter Gale, Rachel Sklar, Renata Quintini, Sarah Ross, Sarah Kunst, Sarah Lefton, Shira Lazar, Sophia Rossi, Stephanie Agresta, Susan Lyne, Tina Seelig, Tina Sharkey, Ty Texidor—the list just goes on and on. This is truly our time, ladies!

Thank you to Soleil Moon Frye for walking me through the whole book process and always being so open and honest about everything—I honestly think of you as a soul sister.

Thank you to Nicole Lapin, both for housing me in NYC so many times, so I could work on this book, and for constantly pushing for me to be on-air, before anybody really cared. You have always been such an incredible champion, I don't know how I can ever repay your amazing kindness and generosity.

Thank you to Erin Kanaley for being such an awesome sidekick and partner in crime for so many years. We had some crazy adventures, girl, didn't we?

Thank you to Peter Jacobs, Amie Yavor, Zach Nadler, Tiffany Chi, and the CAA speaking team for sending me to so many random parts of the world over the past few years to speak. It was during those twenty-four-hour trips to New Zealand, Oman, and Vorbeck, Germany, that the idea for *Dot Complicated* was born, and my United Global Services status accrued. . . .

Special thanks to the übertalented agents at William Morris Endeavor, Jay Mandel, Margaret Riley, Bethany Dick, Miles Gidaly, and Mark Mullet. Can't wait to do amazing things together in the coming years.

Thank you to the incredible team at JonesWorks PR, especially Stephanie Jones, Emily Hofstetter, and Kirby Allison. #awesome #rockstars #trekkingthroughpigpens

A very special thank-you to my good friend Dex Torricke-Barton, one of Silicon Valley's best-kept secrets, for the constant support throughout the writing process. This book wouldn't have been possible without you. I also want to thank Matthew Miller for contributing great ideas and research.

Thank you to the entire Zuckerberg Media team—past, present, and future—Ashmi Pathela, Bradley Lautenbach, Elvina Beck, Emma Paye, Erin Kanaley-Famularo, Holly Leonard, Jeff Paik, Liz Wassmann, Matt Hicks, Niloofar Mansourian, Monica Chambers, Nate Hess, and Ross Siegel. Many, many thanks to all our wonderful investors who have supported our journey.

Thank you to Jeff Paik for always being there for me. From sociology term papers to clearing out dorm rooms, Barbies and all, to dressing in music note caps and singing in the T station, it's been an honor having you by my side.

Thank you to Bradley Lautenbach, my partner since back at the ABC News–Facebook debate! Six awesome years working together and counting. Every single day, I thank my lucky stars that we work together.

Thank you to the most incredible "urban family" a girl could ask for—Chris Kelly, Jen Carrico, Gregg Delman, Becca Schapiro, Kim Lembo, Shari and Jesse Flowers, Rachel Masters, Margot Kaminski, Eric "EZ" Zawid, and Alex and Reina Rampell—we have been together through the good and the bad, the epically wonderful and the truly hideous. From Mexican-themed seders to Jersey Shore–themed going-away parties, crazy Vegas getaways to even crazier Tokyo getaways, toddlers to tiaras, you guys are truly the family you choose.

Thank you to my amazing and supportive real-life family. Edward, Karen, Mark, Donna, and Arielle Zuckerberg, Priscilla Chan, Harry and Jonah Schmidt, Marla and Eron Tworetzky, Grandma Gert, Beast and Luna, and all my extended family in Florida, California, Pennsylvania, and beyond—I am so grateful for your love and support. You have cooked for me, employed me, invested in me (both figuratively and literally), dressed up in Star Wars costumes for me, cheered for me, supported my dreams, and put up with so many of my zany ideas and schemes over the years. I truly wish my grandparents Miriam, Jack, and Sidney could have been here to see me realize my dream of becoming a published author, but I know that they are with me in spirit.

Because even saying thank you twice isn't enough, a heartfelt thank-you to my incredible parents, Edward and Karen Zuckerberg. All those notes in my lunch boxes, those cheers from the front row at countless plays and concerts, those hours in the car—you have always been there for me, always supportive, always loving. You have given me so much, and I am so eternally grateful. Seeing your love extend to my son is the most precious gift you have given me by far.

Thank you to my beautiful son, Asher, who has taught me so many profound and valuable life lessons at such an early age. I hope you never lose your spirit, your zest for life, your desire to sing, your

love of Booney (okay, maybe I won't be crushed if you lose that)—
and may you always live up to the meaning of your name, full of joy
and blessings.

And of course, the biggest thank-you of all goes to Brent
Tworetzky. A loving husband, a wonderful father, and my very best
friend. In the wise words of Toto, "I bless the rains down in Africa"
for you, every single day.

#yay #yourestillreading #thanks #ihearthashtags #icanhazhashtag
#peopleoverusehashtags #waytoomuch #doesanybodyactuallyread
these #whyareyoustillreadingthese #whoooooo #springbreak
#himom #hiharpercollins #wordcountfiller #yayihitmywordcount
#idliketothanktheacademy #pulitzerprize

dot **complicated**

Can't get enough *Dot Complicated*?
Join us online.

The Dot Complicated online community features informative articles on tech trends, etiquette, and more for simplifying and untangling your wired life. Through in-depth articles, practical advice, and quick hacks, Randi Zuckerberg explores the obstacles and opportunities presented by our new online reality.

Visit **www.dotcomplicated.co** and check out some of our past article topics below.

- 10 Great Places to Sell Your Stuff Online

- How to Reach In-box Zero and Stay There

- I'm Scared the Internet Is Ruining My Children

- The 2:00 A.M. E-mail: A Public Display of Ambition

- 5 Things You're Posting Online That Aren't as Harmless as You Think

- 6 Tips for Protecting Your Online Reputation